# RECENT ADVANCES IN
## PLANT NUTRITION

### Volume 1

Proceedings of previous Colloquia on Plant Analysis
and Fertilizer Problems

I.    P. PREVOT, Ed., Institut de Recherches pour les
      Huiles et Oléagineux (IRHO), Paris, 1954,
      263pp.


II.   P. PREVOT, Ed., Institut de Recherches pour les
      Huiles et Oléagineux (IRHO), Paris, 1956,
      410pp.


III.  W. REUTHER, Ed., American Institute of Biological
      Sciences, Washington, D.C., 1961, 454pp.


IV.   C. BOULD, P. PREVOT, J.R. MAGNESS, Eds., American
      Society of Horticultural Sciences, W.F.
      Humphrey Press Inc., Geneva, N.Y., 1964, 430pp.


V.    Abstracts as appendix to Proceedings of VI (this
      volume).


VI.   R.M. SAMISH, Ed., Gordon and Breach Science Pub-
      lishers Inc., New York, N.Y., 1971

# Recent Advances in
# PLANT NUTRITION

*Editor*
R.M. SAMISH

VOLUME 1

GORDON AND BREACH SCIENCE PUBLISHERS

New York      London      Paris

ISBN (Vol. 1):  0 677 12360 4

# CONTENTS OF VOLUME 1

SESSION I.  The Use of Inorganic Tissue Analysis
            for the Determination of the Nutritional
            Status of Plants

## SESSION II. Biochemical Approaches in the Study of the Nutritional Status of Plants

SESSION III. Evaluation of the Nutritional
Potential of the Soil

# CONTENTS OF VOLUME 2

*Continued....*

# SESSION VI.  The Effect of Environmental Conditions on the Nutrient Requirements of Plants

# INTERNATIONAL COMMITTEE ON PLANT ANALYSIS
## AND FERTILIZER PROBLEMS

*President:* Prof. C. Bould, England
*Vice-President:* Prof. J.A. Cook, California
*Organizing Secretary:* Prof. R.M. Samish, Israel

*Past Presidents:*

Prof. J.G. Lundegardh,
  Sweden
Prof. R. Maume, France

Dr. P. Prevot, France
Prof. W. Reuther,
  California

*Members:*

Dr. E.G. Bollard, New
  Zealand
Prof. H.F. Clements,
  Hawaii
Dr. Y. Coïc, France
Prof. F.S. Howlett, Ohio
Mr. J.F. Levy, France
Prof. E. Malavolta,
  Brazil
Dr. P.L. Mitchell,
  Scotland

Prof. J.M. Nielson,
  Denmark
Dr. J.V. Possingham, S.
  Australia
Prof. W. Reuther,
  California
Dr. G. Samuels, Puerto
  Rico
Prof. R. Sato, Japan
Dr. P.F. Smith, Florida
Dr. K.A. Sund, Iran

## ORGANIZING COMMITTEE IN ISRAEL

Dr. Y. Avnimelech, Technion

Dr. A. Bar-Akiva, Volcani Institute

Dr. B. Bravdo (secretary), Hebrew University

Mr. T. Gans, Fert. & Chem. Devel. Council

Dr. H. Lips, Negev Institute

Dr. N. Lahav, Hebrew University

Prof. R.M. Samish (chairman), Volcani Institute & Hebrew University

Prof. Y. Vaadia, Volcani Institute

## SPONSORING INSTITUTIONS

XVIIIth Intern. Horticultural Congress
Israel Academy of Sciences
Ministry of Agriculture
Volcani Institute of Agricultural Research
Hebrew University, Faculty of Agriculture
City of Tel Aviv - Yaffo
Fertilizer and Chemical Development Council and
   Member Companies:
      Dead Sea Works Ltd. (producers of potash and
         bromine).
      Chemicals and Phosphates Ltd. (Producers of phos-
         phate rock and complex fertilizers).
      Haifa Chemicals Ltd. (producers of potassium ni-
         trate and phosphoric acid).
      Arad Chemical Industries Ltd. (producers of phos-
         phoric acid).

# FOREWORD

The Proceedings of this Sixth meeting of the Colloquium use the subtitle "Recent Advances in Plant Nutrition" in order to characterize two approaches which have been emphasized to a greater extent than in previous meetings. First, emphasis has been placed on principles, rather than on mere experimental results. Fertilizer trials, or foliar analytical surveys are often primarily of local nature and therefore can be more appropriately presented in regional meetings. Second, when dealing with fundamentals, the field widens to the entire range of plants and has to include soil-plant relationships. Thus these Proceedings reflect the goal designated by President C. Bould to create a common ground for the meeting of men who work in different sciences and on different crops towards one common aspect of applied plant science.

In addition to the publication of the papers reported at the meeting, this volume also attempts to contribute to an important function of any scientific meeting - the establishment of mutual contacts. It therefore devotes considerable space to discussions and also - a rather unconventional step - publishes photographs and biographies of the actual speakers (not necessarily the senior authors). The address list of participants is designed to help in perpetuating these contacts.

The editors of the Proceedings have limited themselves to the technical aspects of the work and did not feel justified in doing any scientific editing. After all, these papers have already been presented and criticized at the meeting.

Finally it may be appropriate to mention here the major resolutions adopted at the executive meeting of the International Committee for the International Colloquium on Plant Analysis and Fertilizer Problems on March 17, 1970. It approved the direction of development which the Colloquium had taken under the leadership of the outgoing President, Professor C. Bould, and has elected Professor R.M. Samish for the next term. It was decided that the next meeting would be held in the fall of 1974. Readers of the Proceedings of the Sixth Colloquium who find interest in this more fundamental approach to our very practical problems are advised to contact members of the International Committee immediately in order to make possible an even more successful Seventh Colloquium on Plant Analysis and Fertilizer Problems.


R.M. SAMISH

Organizing Secretary

Prof. Cyril Bould, President of the Colloquium

CYRIL BOULD, B.Sc.(Hortic.), M.Sc., Ph.D.(Agric.Chem.)
President of the Sixth Colloquium


Dr. C. Bould was born in Staffordshire, England in
1913 and was educated at Brewood Grammar School and
Reading University. After graduating in Horticulture
he studied the nutrition of narcissus and tulip bulbs
for his Masters degree. From Reading he went to University College of North Wales, Bangor, as Research As-
sistant to the late Professor G.W. Robinson, F.R.S.,
where he carried out research on the chemistry and
fertility of soils in North Wales.

In 1940 he returned to Reading University to carry
out chemical and biological investigations for the Agricultural Research Council on the nature and utilization
of organic waste products for manurial purposes, for
which he was awarded the degree of Ph.D. (Agric.Chem.).

After the war, he was appointed Research Chemist
in the Soils and Plant Nutrition Section, Long Ashton
Research Station, University of Bristol, under the direction of the late Professor T. Wallace, F.R.S., (a
founder member of the International Committee for Plant
Analysis and Fertilizer Problems). Here he developed
his interest in plant nutrition, specialising in the
nutrition of fruit crops.

His main research contributions have been concerned with the effect of cover crops on tree nutrition
and in developing leaf analysis (foliar diagnosis) as a
guide to the nutrition of soft fruits.

In 1952 he became Head of the Nutrition of Fruit
Plants Section at Long Ashton Research Station.

When Professor Wallace retired in 1956, Dr. Bould
took his place as a member of the International Committee on Plant Analysis and Fertilizer Problems. He
was Organizing Secretary of the 4th International Colloquium (Brussels, 1962) and Vice-President (Acting
President) of the 5th Colloquium in Maryland, U.S.A.,

in 1966.

He has travelled widely, visiting research institutes in Europe, Africa and North America. In 1967 he was Visiting Professor of Horticulture at Michigan State University where he gave a postgraduate course of lectures on plant nutrition.

Since 1969 he has been seconded by the British Ministry of Overseas Development to Kenya as Director of Research of the Coffee Research Foundation, Ruiru.

Dr. Bould is married and has two daughters.

PRESIDENTIAL ADDRESS

It is my pleasure and privilege to welcome you to
the 6th International Colloquium on Plant Analysis and
Fertilizer Problems, and to wish you all a pleasant and
useful few days of discussion and talk, because that
is what a colloquium is for; it is an assembly for
discussion, *not* a platform for reading papers and presenting data. I would ask you to support the Organizing Secretary's plea to contributors: :Let us get together and discuss new ideas rather than hear additional
facts supporting accepted principles!"

I would like to spend a few moments recalling the
origin and history of these Colloquia. Although I was
not present at the first Colloquium I knew a number of
the founder members. At that time crop nutritionists
had no suitable representative body or forum for discussion. The scientific disciplines of Botany, Chemistry, Physics and Soil Science all had established
international bodies for organizing meetings and con-
gresses, but plant nutrition, which makes use of these
disciplines, had no such organization. So a number of
leading research workers in this field, including
Lundegårdh (Plant Physiology), Wallace (Soil Science
and Fruit Nutrition), Prevot (Physiology), Steenbjerg
(Soil Fertility), Reuther (Nutrition) and Mitchell
(Spectrochemistry), to mention but a few, decided to
meet on the occasion of an appropriate International
Science Congress to discuss how their various disciplines and expertise knowledge could help to explain,
or solve, many complex soil-plant-nutritional problems.

About that time new chemical analytical techniques
and sensitive instruments (e.g. spectrograph) were being developed for soil and plant analysis. As a result of these developments, plant analysis was being
used increasingly for the diagnosis of nutrient deficiencies, and data from leaf analysis were becoming

accepted as an index of plant nutritional status and as a guide to fertilizer application.

The first Colloquium was held on the occasion of the International Botanical Congress in Paris in 1954 under the title of "Plant Analysis and Fertilizer Problems." It was organized by Dr. P. Prevot and was sponsored by L'Institut de Recherches pour les Huiles et Oleágineux (I.R.H.O.). Further Colloquia, under the same title were held in Montreal (1959), Brussels (1962) and Maryland (1966).

What progress has been made over this period in our understanding of crop nutrition in the field, and what should be our future objectives?

From a practical point of view the grower needs reliable advice on a number of important questions. First, how can he assess the current nutritional status of his crop; secondly, what nutritional level is required at specific stages of growth and development for economic optimal yields of high quality produce; thirdly, what is the relationship between soil nutrient content and plant nutrition status and fourthly, at any given level of soil fertility what economic quantities of applied nutrients are required (and when) to raise the plant nutrient level from deficiency to sufficiency.

If we could provide answers to all these questions, then maximum crop potential for any given environment could be attained (subject to adequate control of pests and diseases). But we are not yet in a position to give accurate, reliable advice on all these questions for all crops on all soil types.

Given the necessary facilities, however, it is now possible to diagnose accurately all nutrient deficiencies and toxicities by using a combination of visible symptoms (visual diagnosis), biochemical indicators, physiological responses and plant analysis (root, petiole, leaf or fruit). Roots are often best for diagnosing heavy metal toxicities, petioles and leaves for N, P, K and Mg and for Cu, Zn, Mn and Fe deficiencies and fruits for Ca and B deficiencies. Enzyme activity, or the accumulation of metabolic products, may also be useful for diagnosing sub-clinical deficiencies.

Accurate diagnosis is the starting point for sound advice. A knowledge of "critical," sufficient or optimal nutrient levels and ratios in plant organs at different physiological growth stages is the ultimate target. By using a combination of sand and water cultures and field experiments, based on factorial and response surface designs, the nutrient concentration ranges associated with deficiency, sub-clinical deficiency, sufficiency and excess have been determined for many crops.

The weakest link is the advisory chain is a lack of knowledge concerning the relationships between soil nutrient content and plant nutrient uptake, and of the responses - as measured in terms of plant growth, crop yield and changes in plant nutrient status - to applied nutrients on different soil types and differing levels of soil fertility. On consideration this is not surprising.

Different crops explore quite different amounts of the total soil volume, estimated at from 1 to 5% in the field and more if the root-run is confined. Furthermore, the concentration of soil nutrients available at any one time, for uptake by the roots, is determined by such factors as soil moisture content, mass-flow and diffusion, by the nature of the soil colloids and by their relative contents of anions and cations. For those plant nutrients that are readily fixed by the soil, placement in relation to root activity is important for maximum uptake and crop response.

It seems to me that in future much greater emphasis and more research is needed on soil-plant

nutrient relationships.   In addition to fundamental
studies on the mechanisms of nutrient uptake, movement
and function, applied studies are needed to provide an
*extensive body of information for each crop* and for
*each major soil type* on (1) the relationship between
available soil nutrient levels and plant nutrient contents and (2) on the responses to applied nutrients on
these soils as measured by enhanced plant growth and
by changes in plant nutrient status.

   With these few remarks I have pleasure in opening
the 6th Colloquium for the presentation of papers and
for discussion.


                                    CYRIL BOULD
                                    *President*

Opening the Colloquium

The Colloquium in session. × Dr. P. Prevot, founder of the Colloquium

# THE USE OF INORGANIC TISSUE ANALYSIS
## FOR THE DETERMINATION OF THE
## NUTRITIONAL STATUS OF PLANTS

Session Leader:  CYRIL BOULD

# A MEANING FOR FOLIAR DIAGNOSIS

Euripides Malavolta[*]

*Department of Chemistry, E.S.A. "Luiz de Queiroz," University of Sao Paulo, Piracicaba, São Paulo, Brazil*

Vivaldo F. Da Cruz

*Department of Maths. and Statistics, Escola Superior de Agriculture "Luiz de Queiroz," University of São Paulo, Piracicaba*

ABSTRACT

Foliar diagnosis is considered here as a method for evaluating soil fertility. This is based on the assumption that correlations do exist between the dosage of fertilizer and both yield and level of the corresponding element in the leaves.

There are three important steps in foliar diagnosis: establishment of the concept as such, experimental work, and extension of the results to the farmer. This paper describes a particular application of these steps.

A somewhat new definition of the critical level is presented, which takes into account the questions that must be answered when making fertilizer recommendations, namely: what kind, how much, and will it be profitable?

[*] Euripides MALAVOLTA. Prof. Biochem., E.S.A. "Luiz de Queiroz" Univ. S. Paulo, Piracicaba, S. Paulo (Brazil). b. 1926 Araraquara (S. Paulo); 1948 B.Sc. (Agric. Sci.) and 1951 D.Sc. (Agric. Chem.), E.S.A. "Luiz de Queiroz" Univ., S. Paulo (Brazil).

1

The critical level is defined as the range of a given element in the leaf, below which the yield is limited and above which the use of fertilizer is no longer economical. Field trials made it possible to show that the well-known Mitscherlich equation fits the experimental data reasonably well, that is:

$$y = A(1 - 10^{-c(x + b)})$$  (1)

where y is the yield due to x - amount of element supplied plus b - the soil supply; c is the coefficient of carrier efficiency and A is the maximum yield. The values of the parameters A, b and c could therefore be calculated.

The dosage of element capable of maximum economical yield was calculated using the equation

$$x^* = 1/2x_u + (1/c) \log \frac{wu}{tx_u}$$  (2)

where $x^*$ is the dosage of element giving the maximum economical yield, $x_u$ is the dosage of element which caused u = the increase in yield relative to treatment without the element; w is the unit price of the agricultural product and t = the unit price of the fertilizer element.

Analysis of samples collected at the appropriate time has shown that the following types of relationships usually hold:

$$Y = a + dx$$
or $$Y = a + dx + ex^2$$  (3)
or $$Y = a + dx - ex^2$$

where Y is the level of element in the leaves corresponding to x, the dosage of element; d and e are coefficients.

The critical level, as defined above, is then calculated by setting x = x* in the regression equations given in (3).

The advantages and limitations of the procedures are discussed and illustrated.

## 1 THEORY

Foliar diagnosis as a method for assessing the fertilizer requirements of a given crop is based on the assumption that, within certain limits, there is a positive correlation among doses of nutrient supplied, leaf content of this element, and yield. This method of studying problems of soil fertility involves the use of the plant itself as an extracting agent for its nutrients.

The recommendations for fertilizer use should answer the following questions: a) What kind? b) How much? c) When? d) How? e) Will it pay?

There is, obviously, no one simple answer to such a complex series of questions. It was attractive, however, to envisage one single approach to questions a, b and e. To do so, the established "critical level" concepts were modified by the introduction of an economics component: "the level of a given element in the leaf below which yield is limited and beyond which the use of fertilizer is no longer economical" (Malavolta & Pimentel Gomes, 1961; Malavolta et al., 1962). This physiological-economical critical level could be calculated through the following steps:

By conducting a sufficient number of field trials it would be possible to calculate the parameters A,b, and c of the Mitscherlich equation, whenever it would fit the experimental data;

The dosage of an element capable of giving the maximum economical yield would be calculated with the aid of the equation:

$$x^* = 1/2 \ x_u + (1/c) \ \log \frac{wu}{tx_u}$$

where $x^*$ = dosage of element which gives the maximum economical yield, $x_u$ = dosage of element which caused u = increase in yield in relation to the treatment without the element; w = unit price of the agricultural product and t = unit price of fertilizer element.

By substituting $x^*$ in the regression equations defining the relationship between fertilizer applied dosage and leaf element level, the value of the critical level Yc would be found.

The extension of the method to the evaluation of fertilizer needs under similar soil and climate conditions would finally depend upon the introduction of the observed level of element in the corresponding equation.

The applicability of the concept was demonstrated as described in the following sections.


## 2 EXPERIMENTAL


Forty NPK 3×3×3 factorial experiments were carried out in the several sugar-cane growing areas in the State of São Paulo, Brazil. Nitrogen was supplied at the rates of 0, 60 and 120 kg per hectare; both phosphorus and potassium were applied at the rates of 0, 75, and 150 kg per hectare. For analysis, the third and fourth leaves from the top were collected when the plants were 4-5 months old, the midrib being discarded; only the middle third of the green tissue was used for chemical determination of the total nitrogen, phosphorus and

TABLE 2.1.  Most Economical Dosages of Elements (x*)
and Critical Levels (Yc)

| Element | x* kg/ha | Yc % |
|---|---|---|
| Nitrogen (CN) | 93 | 1.94 |
| Phosphorus ($P_2O_5$) | 24 | 0.17 |
| Potassium ($K_2O$) | 155 | 1.62 |

TABLE 2.2.  Comparison of Maximum Economical Yields
(yc) and Critical Levels (Yc) Calculated and Found
Experimentally

| Effect of | yc t/ha | | Yc % | |
|---|---|---|---|---|
| | calculated | observed | calculated | observed |
| Nitrogen | 126 | 115 | 1.93 | 2.00 |
| Phosphorus | 129 | 124 | 0.17 | 0.16 |
| Potassium | 131 | 129 | 1.60 | 1.50 |

potassium.  For the statistical analysis of yield and
chemical data, the corresponding results were grouped
according to soil type.  In the particular case of the
fifteen experiments located in "terra roxa misturada,"
the values for x* and critical levels are given in
Table 2.1; calculations were based on the price of
ammonium sulfate, simple superphosphate and potassium
chloride.

The regression equations defining the relationship
between element applied (x) and its concentration with
leaves (Y) were found to be:

for nitrogen      $Y = 1.808+0.00249x-0.000011x^2$;

for phosphorus    $Y = 0.1669+0.0001051x-0.0000004032x^2$;

for potassium     $Y = 1.35+0.002041x-0.000001778x^2$.

A series of 12 experiments were set up next.   In
4 of these, 5 levels of nitrogen were supplied, the
other two elements being given in non-limiting amounts;
in the next four experiments, the same was done with
respect to phosphorus, and in the remainder potassium
was the variable.  By having doses of elements below
and above $x*$ and by analyzing leaf samples it would be
possible to check the validity of the calculated
values for economical doses of element and the
critical level.  Table 2.2 summarizes the results
obtained.  It seems that a reasonable agreement
between calculated and observed values was achieved.

## 3 EXTENSION

A homogenous plot planted with 4.5 month-old
sugar cane was split into 8 sub-plots.  Out of it 4
were selected at random.  In each one 6 ratoons were
numbered also at random; in each ratoon 4 stools were
selected; the third and fourth leaves from 2 stools
were collected for analysis as previously described,
total nitrogen, phosphorus and potassium being determined.  Statistical analyses summarized in Table 3.3
show that minimum variance is obtained when 20 stools
constitute one sample, each stool corresponding to one
ratoon.

The sampling technique thus ascertained was used
for the survey of the nutritional status of sugar cane
growing in 10 soil series in the county of Piracicaba,
S. Paulo, one of the main sugar-producing areas in

Brazil.  10 samples were collected in areas of less
than 5 hectares.

TABLE 3.1  Mean and Standard Deviation from the Mean

| Variable | Nitrogen | Phosphorus | Potassium |
|----------|----------|------------|-----------|
| $\bar{x}$ | 1.96 | 0.18 | 2.58 |
| $s(\bar{x}_{20})$ | 0.04695 | 0.00869 | 0.21222 |
| $s(\bar{x}_1)$ | 0.02433 | 0.00237 | 0.02333 |

Leaf values, Ya were interpolated in the regression equations given in Section 2, which per-
mitted the calculation of $x_a$ and therefore the
determination of $x^* - x_a$ which gives the rate of
element to be applied to obtain the maximum economical
yield.  Results thereof are given in Table 3.2.
Critical values were recalculated to adjust tor
changes in price of cane and fertilizer, being 1.94%
for N; 0.17% for P; 1.62% for K.


4 SUMMARY AND CONCLUSIONS


In this paper foliar diagnosis is considered as a
method designed to evaluate soil fertility, that is,
the quantitative need for fertilizer through the
chemical analyses of leaves.  It is based on the
assumption that, within limits, positive and significant correlations do exist between the following
variables: dosage of fertilizer applied and yield;
dosage of fertilizer and level of corresponding element in the leaves.

The development of a useful concept for foliar

TABLE 3.2. Leaf Values and Most Economical Rates of Elements

| Soil Series | Nitrogen | | Phosphorus | | Potassium | |
|---|---|---|---|---|---|---|
| | Ya % | x* - ya kg/ha | Ya % | x* - xa kg/ha | Ya % | x* - xa kg/ha |
| Tanquinho | 1.53 | 175 | 0.18 | --- | 1.57 | 29 |
| Luiz de Queiroz | 1.12 | 197 | 0.14 | 159 | 1.17 | 228 |
| Dois Carregos | 1.84 | 77 | 0.20 | --- | 0.56 | 454 |
| Iracema | 1.60 | 156 | 0.19 | --- | 0.93 | 326 |
| Quebra Dente | 1.36 | 210 | 0.19 | --- | 1.60 | 12 |
| Ibitiruna | 1.52 | 176 | 0.19 | --- | 1.87 | --- |
| Saltinho | 1.29 | 223 | 0.17 | --- | 1.19 | 220 |
| Godinho | 1.58 | 162 | 0.21 | --- | 1.49 | 77 |
| Sertãozinho | 1.48 | 184 | 0.16 | 89 | 1.18 | 222 |
| Litosol | 1.36 | 211 | 0.16 | 44 | 1.51 | 59 |

diagnosis requires three steps, other points (such as sampling) to be clarified beforehand: (a) establishment of the concept as such; (b) demonstration of its validity through experimental work; (c) extension of the results to the farmer.

This paper discusses how this was achieved in a particular case.

The concept advanced dealt with a somewhat new definition of critical levels which took into consideration the questions that must be answered when making recommendations: (a) what kind? (b) how much? and (c) will it pay?

The critical level was defined as the range of a given element in the leaf below which yield is limited and above which the use of fertilizers is no longer economical.

By conducting a sufficient number of field trials it was possible to show that the well known Mitscherlich equation fits the experimental data reasonably well, that is

$$y = A \left[ 1 - 10^{-c(x + b)} \right]$$

where y = yield due to x = amount of element applied plus b = soil supply. c = coefficient of efficiency of the carrier and A = maximum yield. The values of the parameters A, b and c could thus be calculated.

The dosage of element capable of giving the maximum economical yield was calculated with the aid of the equation

$$x^* = 1/2\, x_u + (1/c) \log \frac{wu}{tx_u}$$

where $x^*$ = dosage of element which gives the maximum economical yield; $x_u$ = dosage of element which caused

u = increase in yield in relation to the treatment
without the element; w = unit price of the agricultural product and t = unit price of fertilizer element.

The analysis of samples collected at the appropriate time has shown that the following types of
relationship usually hold:

$$Y = a + dx$$

or $$Y = a + dx + ex^2$$

or $$Y = a + dx - ex^2$$

where Y = level of element in the leaves corresponding
to x = dosage of element. d and e = coefficients.

The critical level as defined in 5 is then calculated by making x = x* in the regression equations
given in this section.

The physiological-economical concept of critical
level was tested in several experiments with sugar
cane with satisfactory results. The validity of the
dosage recommendations was further checked in another
series of field trials.

A preliminary survey of the nutritional status of
sugar cane in a small region was conducted after
determining the minimum number of leaf samples which
represents a given area; the results thereof are now
being evaluated in order to permit its diffusion to
the sugar cane planters.

ACKNOWLEDGEMENTS

The work described in this paper was carried out
with the help of the following institutions: the
Rockefeller Foundation, New York, U.S.A.; Conselho
Nacional de Pesquisas, Rio de Janeiro, Gb., Brazil;
Fundação de Amparo à Pesquisa do Estado de São Paulo,
S.P., Brazil.

REFERENCES

1. MALAVOLTA, E. and PIMENTEL GOMES, F. (1961) Foliar
   diagnosis in Brazil. *Amer. Inst. Biol. Sci. Publ.*
   *8*: 180-189.

2. MALAVOLTA, E., PIMENTEL GOMES, F. and COURY, T.
   (1963) Foliar diagnosis as a tool for assessing
   fertilizer needs in developing countries. *U.N.
   Conf. Appl. Sci. Techn. Benefit Less Developed
   Areas. Doc. 39/c185*, 4 pages.

*Questions to Prof. Malavolta*


CLEMENTS: How much sunlight is available to your crop
as well as temperatures?

MALAVOLTA: These belong to the limitations of this method.
We have no respective data·in Brazil. There are some
which were collected in South America by Evans, which
show that they could cause quite a difference. We tried
to obviate these factors by collecting the samples under
similar climatic conditions as those which prevailed
when the method was standardized.

   We are trying to get some rough estimates of fertilizer needs of sugar cane using an annual diagnosis, and
later on we are going to refine this method as much as
we can.

AMITAI COHEN: Did you take into consideration the influence of nitrogen on percentage and total yield of
sugar?

MALAVOLTA: We have confirmed the known fact that sometimes a high nitrogen application diminishes the amount
of sugar produced.

ALÈ:  Est-ce-que vous vous servez des facteurs de
correction, une pour la variété, une pour l'âge de la
canne, et une correction pour l'état indirect des
tissus?

MALAVOLTA:  The correction for variety is rather small
because it is pretty much standardized in the region
where I have worked.  We do have correction factors for
age of the plants, when the sampling procedure is not
carried out according to the standards established previously.  But we try, as much as we can, to collect
samples for diagnostic purposes when the plants are from
4 to 5 months old.  At that stage the leaves are fully
matured and there is no die-back in the leaves which are
collected as samples.

WEHRMAN:  The cost of fertilization would be expected to
form only a few percent of the total production costs;
how much does it actually amount to?

MALAVOLTA:  It is about 25% of the total cost of production, when distribution is carried out by machinery.

CLEMENTS:  With us, it is less than 4 or 5%; if we apply
the fertilizer by hand, the calculation would be based
on the poundage applied, and that is relatively low.
If we are putting on a fairly heavy amount, strangely
enough, it is fairly cheap to do it by hand.  But if we
have a very light amount to put on, we do it by air-
plane, which brings it to about 2 dollars an acre.

    If we do it by a machine, then it may be between 2
and 3 dollars an acre, depending on the exact equipment used.  As far as the main point of the lecture is
concerned, I must confess that I don't get very excited
about equations where the moisture level and the energy
involved are ignored.

MALAVOLTA: I pointed out that the use of the Mitscherlich equations is one of the limitations of the method. But, there was very good agreement between calculated and observed values. The experience that has been obtained in Brazil so far show that Mitscherlich equations fit the data quite well. There are cases, of course, where other equations have been used.

KAFKAFI: I got the impression that you have in these equations three unknown variables. You used three fertilizer levels to fit a Mitscherlich equation. If you use three, then equations will always go perfectly through the three points. Is it still possible to apply the fertilizer by machines, after you have decided by leaf analysis that you have any deficiencies?

MALAVOLTA: It is still possible when the sugar cane is from four to six months old; when the plants are older, then the crop closes in and it is impossible, or nearly impossible, to distribute fertilizer by machinery. Then they have to use airplanes, or to distribute some fertilizers in the irrigation water.

NIELSEN: Do you have to use those equations at all, when a man has been given such wonderful ability with his hands and his spirit, to make a smooth curve?

MALAVOLTA: There would be no need for mathematical diagnosis and other complications if the parameters were constant. But since there are variations from soil to soil, the dosages of fertilizer to be applied have to vary. And the only way that has been found to do this in Brazil is by using those equations, which take into consideration the actual nutrient status of the plant as it relates to foliar diagnosis.

CARY: You mentioned certain levels of super phosphate applications. Are they in terms of P?

MALAVOLTA: It is $P_2O_5$.

# THE IRRATIONALITY OF USING LEAF ANALYSIS AS A UNIQUE REFERENCE TO CITRUS FERTILIZER REQUIREMENT

Peter R. Cary

*CSIRO, Division of Irrigation Research, Griffith, New South Wales, Australia*

ABSTRACT

In a factorial experiment with citrus an inherent soil phosphorus deficiency was aggravated by soil acidity caused by cumulative heavy applications of ammonium sulfate applied during 1947-1965. Leaf analysis, however, merely indicated a phosphorus deficiency; superphosphate applications at rates of 400 kg/ha/year were of no avail on plots receiving the highest rate of ammonium sulfate until calcium carbonate was applied at rates almost double that of the total ammonium sulfate applications. Thus, on plots receiving cumulative ammonium sulfate at the rate of 29 t/ha (or 1.6 t/ha/year) calcium carbonate applications were made at the rate of 50 t/ha. This way, phosphorus deficiency was overcome; yields were doubled on some plots, and fruit quality was markedly improved.

More recently, experiments with citrus cuttings, testing effects of root temperature, nutrient supply and crop load on growth, yield and fruit composition, have shown that leaf analytical levels bear little relation to overall productivity when above a critical

Peter R. CARY. Exp. Off., Div. Irrig. Res., CSIRO, Griffith (N.S.W., Australia) since 1964. b. 1927 London; 1954 B.Sc., Univ. London; 1954 Soil Scientist, Nat. Agric. Advisory Serv., Cambridge (Eng.); 1961 Fruit Chemist, Dept. Agric., Yanco, N.S.W.

minimum.  Leaf levels of N and P were relatively unaffected by root temperature and crop load treatments,
despite differences in yield and dry matter production.  Root temperature has been shown to be a
critical factor; at 25°C citrus cuttings were able to
maintain satisfactory growth even at very low or near-
deficient NP nutrient supply, and at 19°C cuttings
were unable to take full advantage of high NP nutrient
supply.


INTRODUCTION


     Too frequently the use of plant or soil analysis
is regarded as the panacea to the farmers' problems.
In evaluation of leaf analytical results Archibald
(1964) has mentioned that such factors as variety,
system of soil management, pruning, spray programs,
time and place of sampling, and weather conditions
should be taken into account.  The need to consider
nutrient concentrations in terms of dry matter production has been mentioned by Steenbjerg (1964).  This
paper is written to stress the importance of relating
analytical results to overall dry matter production,
crop yield and quality factors.


MATERIALS AND METHODS


     Samples of leaf material were obtained from a
long-term citrus factorial field experiment previously
described by Bouma and McIntyre (1963) and by Cary
(1968), and from a sand-culture experiment with citrus
cuttings described by Cary (1970).  Four to five
month-old spring flush leaves were collected in
January 1967 from each plot of the field experiment,
in accordance with the sampling technique described by
Chapman (1960).  Leaf samples from the sand-culture

experiment were obtained when the crop was harvested
in May, 1967.

## Field experiment

Over a period of 19 years (1947-1965) the field
experiment tested the effects of four different rates
of nitrogen (0, 85, 170 and 340 kg/ha, designated $N_0$,
$N_1$, $N_2$ and $N_4$ respectively) applied annually as
ammonium sulfate.   Two tillage and two non-tillage
treatments also were tested.   Tillage involved growing
a cover crop from March to September, and maintaining
the ground in a clean-cultivated state during the remainder of the year.   A cover crop of tick beans and
oats was grown on tillage treatment $C_1$, and subterranean clover on treatment $C_2$.   Non-tillage treatment $C_3$ involved the growth of a volunteer cover crop
which was kept well mown.   A completely bare soil
surface was maintained on non-tillage treatment $C_4$ by
spraying the ground with power kerosene about four
times a year.   Since 1955 superphosphate was applied
annually at the rate of 37 kg P/ha.   The experiment
was concluded in 1965, and results have been reported
by Cary (1968); for the period 1961-1965 the highest
yields of the best quality fruit were obtained from
the $N_1C_4$ treatment.   Since 1965 remedial nitrogen and
calcium treatments have been applied to restore the
experimental area to a uniformly high level of productivity.

## Sand-culture experiment

In mid-December 1965 shoots from vigorous
Washington Navel orange trees with well developed
spring flush leaves were selected and mist-propagated
(Cary, 1970).   In October 1966  the cuttings flowered
and set fruit.   Three crop-load (0, 1 and 2 fruits
per cutting), two root-temperature (19 and 25°C) and
four nutritional treatments ($N_1P_1$, $N_1P_2$, $N_2P_1$ and

$N_2P_2$) were imposed.  Nutrient treatments were applied
daily.  $N_1$ and $P_1$ treatments corresponded to near-
deficient or subnormal levels of nitrogen and phosphorus respectively, and $N_2$ and $P_2$ to near-luxury
levels.

RESULTS

*Field experiment*

In Table 1 the effects of nitrogen and cultural
treatments on yields and juice contents for the four-
year period 1965-1968 are shown.  Remedial nitrogen
and calcium treatments, applied since 1965, did not
have any substantial effects on yield and quality until 1967.  Consequently, the two periods 1965-1966
and 1967-1968 can be considered as representing conditions before and after the application of remedial
treatments.  In Table 1 average yields are given for
these two periods in terms of edible matter (yield x%
juice content) for each cultural and nitrogen treatment. Juice contents, and leaf N, P and K values as of
January 1967, are included for comparison purposes.
The 1967-1968 results indicate that remedial treatments had smoothed out former yield differences between nitrogen treatments to a marked degree; the $N_4$-
treated trees, however, still gave lower yields and
lower juice contents.  Results also indicate that,
despite low leaf nitrogen levels (< 2.0% on a leaf dry
matter basis), high yields were obtained in 1967 and
1968 from $N_0$ and $N_1$-treated trees.  Furthermore, the
fruit was of a higher quality than from $N_4$-treated
trees showing a more adequate leaf nitrogen status
(> 2.0%).

Since responses to calcium treatments occurred
mainly on tilled ($C_1$ and $C_2$) plots receiving $N_2$ and
$N_4$ nitrogen treatments, results in Table 1 are ex-

TABLE 1

Effects of Nitrogen and Cultural Treatments on Yield of Edible Matter
(Yield x% Juice Content) and on Juice Content, Expressed as the Average
of Two Periods Each of Two Years Duration. Results are Compared with
Leaf N, P and K Levels as of January, 1967.

| Nitrogen* or cultural** treatment | Yield of edible matter (t/ha) | | Juice content (%) | | As % of leaf dry matter | | |
|---|---|---|---|---|---|---|---|
| | 1965-6 | 1967-8 | 1965-6 | 1967-8 | Nitrogen | Phosphorus | Potassium |
| $N_0$ | 8.9 | 19.1 | 40.3 | 48.2 | 1.71 | 0.117 | 1.57 |
| $N_1$ | 12.8 | 19.6 | 39.8 | 46.2 | 1.77 | 0.094 | 1.43 |
| $N_2$ | 11.6 | 19.2 | 37.6 | 44.2 | 1.95 | 0.088 | 1.52 |
| $N_4$ | 7.8 | 16.8 | 34.2 | 41.7 | 2.24 | 0.084 | 1.73 |
| $C_1$ | 8.8 | 16.9 | 37.1 | 44.4 | 1.95 | 0.095 | 1.51 |
| $C_2$ | 10.4 | 18.2 | 38.2 | 44.6 | 2.03 | 0.092 | 1.50 |
| $C_3$ | 10.0 | 17.5 | 37.8 | 45.2 | 1.80 | 0.096 | 1.62 |
| $C_4$ | 11.8 | 22.4 | 38.8 | 46.2 | 1.90 | 0.099 | 1.63 |
| L.S.D. at P = 0.05 | | | | | 0.16 | 0.009 | 0.17 |
| P = 0.01 | | | | | 0.22 | 0.012 | 0.23 |

* $N_0$, $N_1$, $N_2$, $N_4$: Annual applications of ammonium sulphate at the rate of 0, 85, 170, 340 kg N/ha respectively from 1947 to 1965. $N_2$ and $N_4$ plots received no nitrogen in 1966 and 1967, but in 1968 received 100 kg N/ha. $N_0C_4$ plots received 255 kg N/ha in 1966 and 100 kg N/ha in each subsequent year. All other plots received 100 kg N/ha annually from 1966 onwards. Nitrogen was supplied in 1966 and subsequently in the form of an ammonium nitrate limestone fertilizer.

** $C_1$, $C_2$, $C_3$, $C_4$: Winter tick bean, winter clover, permanent sod and bare surface treatments respectively.

pressed as the average of limed and unlimed plots.
Furthermore, responses to liming occurred mainly in
"on" or high yielding years, when, presumably, there
was a greater phosphorus requirement. Bouma (1956 and
1959) found that the low yields and poor fruit quality
evident in the mid-1950's were caused by low phosphorus supply. Consequently, since 1955 superphosphate treatments at an average annual rate of 37 kg
P/ha have been applied to the whole experimental area.
Though there was no evidence to indicate that trees
were suffering from calcium deficiency (Groenewegen
and Bouma, 1960), from 1955 to 1965 small annual
applications of finely-ground calcium carbonate were
applied to nitrogen plots in such a way that half the
plots received lime and the other half no lime.  These
liming treatments (0, 750, 1500 and 3000 kg $CaCO_3$/ha/
year) were made to reduce soil acidity produced by
cumulative ammonium sulfate applications, and thus -
it was hoped - improve phosphorus availability.

Beneficial effects of lime can most conveniently
be assessed from Fig. 1 which compares yield data
since 1955 from limed and unlimed $N_4C_1$-treated plots.
On unlimed $N_4C_1$ plots little benefit was obtained from
superphosphate applications until 1967-1968; mean
annual yields from unlimed $N_4C_1$ plots showed a slight,
but progressive decline over the three four-year
periods 1955-1958, 1959-1962 and 1963-1966.  The increase in 1967-1968, however, indicates a response to
the nil-nitrogen treatments applied to $N_2$ and $N_4$ plots
during 1966 and 1967 (Table 1).  From 1960 to 1966
limed $N_1C_4$ plots showed a small beneficial response to
liming treatments applied annually from 1955 to 1965.
However, yields and quality of fruit from limed $N_4C_1$
plots did not closely approach the levels obtainable
from the optimum $N_4C_1$ treatment until trees had had
sufficient time to benefit from the 20 t/ha calcium
carbonate application made in October 1965.  At this
final stage limed $N_4C_1$ plots had received a total
calcium carbonate application of 50 t/ha; this was

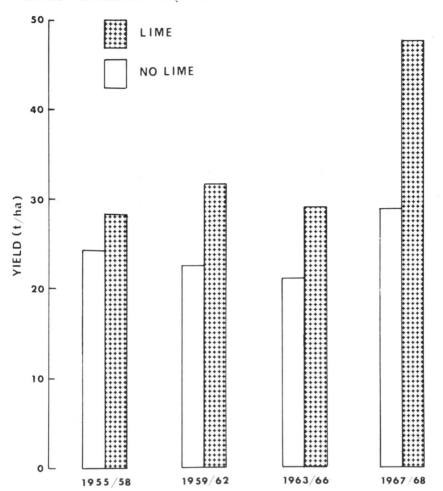

FIG. 1. Average annual yield of citrus as affected by
calcium carbonate applications. Yields are expressed
as the average of three periods (1955-1958, 1959-1962,
1963-1966) each of our years duration, and of one
period (1967-1968) of two years duration. From 1947 to
1968 trees received the $N_4 C_1$ treatment (see Table 1).
Calcium carbonate was applied to limed plots annually
from 1955 to 1965 at 3 t/ha, and finally in October
1965 at 20 t/ha.

## TABLE 2

Effects of Nutrient and Root Temperature Treatments on Dry Matter Production of Washington Navel Orange Cuttings, and on Leaf N and P Status

| Nutrient treatment* | Root temperature (°C) | Weight of dried material (gm.) | | | | Juice content (%) | As % of leaf dry matter | |
|---|---|---|---|---|---|---|---|---|
| | | Roots | Stems | Leaves | Fruit | | Nitrogen | Phosphorus |
| $N_1P_1$ | 19 | 9.0$^a$ | 11.7$^{ab}$ | 25.4$^{ab}$ | 42.2$^a$ | 44.8$^a$ | 2.13$^b$ | 0.079$^a$ |
| | 25 | 16.6$^b$ | 13.6$^{abc}$ | 27.2$^{ab}$ | 43.8$^a$ | 43.2$^a$ | 2.08$^b$ | 0.073$^a$ |
| $N_1P_2$ | 19 | 8.5$^a$ | 10.4$^a$ | 22.0$^a$ | 42.5$^a$ | 47.2$^b$ | 2.07$^b$ | 0.134$^b$ |
| | 25 | 18.5$^b$ | 17.4$^{cd}$ | 33.2$^{bc}$ | 53.7$^b$ | 48.5$^b$ | 1.73$^a$ | 0.159$^b$ |
| $N_2P_1$ | 19 | 10.0$^a$ | 11.5$^{ab}$ | 27.1$^{ab}$ | 40.1$^a$ | 43.5$^a$ | 2.61$^{cd}$ | 0.072$^a$ |
| | 25 | 17.8$^b$ | 15.6$^{bc}$ | 31.1$^{bc}$ | 44.3$^a$ | 44.9$^a$ | 2.55$^{cd}$ | 0.076$^a$ |
| $N_2P_2$ | 19 | 11.0$^a$ | 15.8$^{bc}$ | 38.2$^c$ | 40.8$^a$ | 44.4$^a$ | 2.52$^c$ | 0.142$^b$ |
| | 25 | 18.7$^b$ | 21.8$^d$ | 48.8$^d$ | 55.1$^b$ | 47.7$^b$ | 2.74$^d$ | 0.152$^b$ |

* $N_1P_1$ - $N_2P_2$, as previously described in Materials and Methods. Statistical analysis carried out on logarithmically transformed data. Treatment means in the same column whose postscripts have a letter in common do not differ at P = 0.05 (Duncan, 1955).

almost double the total application of ammonium sulfate  (29 t/ha) made to these plots during 1947 to
1965.

*Sand-culture experiment*

Table 2 shows the effects of nutrient and root
temperature treatments on dry matter production of
citrus cuttings; leaf nitrogen and phosphorus levels
are also included for comparison purposes.

$N_1P_1$-treated cuttings at a root temperature of
19°C produced about half the dry matter production of
$N_2P_2$-treated cuttings at 25°C; these, in turn, produced more dry matter and significantly higher yields
of fruit with higher juice contents than $N_2P_2$-treated
cuttings at 19°C.  Thus, as also indicated by leaf
phosphorus levels, the higher near-optimum root temperature (25°C) appears to be important in facili-
tating phosphorus uptake from the root zone.  Almost
as much dry matter was produced from $N_1P_2$-treated
cuttings at 25°C as from $N_2P_2$-treated cuttings at the
same temperature.  This may indicate that citrus is
more dependent on an adequate phosphorus supply than
on an adequate nitrogen supply.

REFERENCES

1.    ARCHIBALD, J.A. (1964)  Weather effects on leaf-
         nutrient composition of fruit crops.  *Plant
         Analysis and Fertilizer Problems IV*: 1-8.  Bould,
         C., Prevot, P., and Magness, J.R. (eds.).  Amer.
         Soc. Hort. Sci., New York.

2.    BOUMA, D. (1956)  Studies in citrus nutrition.  II:
         Phosphorus deficiency and fruit quality.  *Aust.
         J. Agric. Res.*, *7*: 261-271.

3.    —————— (1959)  Growth, yield and fruit quality
       in a factorial field experiment with citrus in
       relation to changes in phosphorus nutrition.
       *Aust. J. Agric. Res.*, *10*: 41-51.

4.    BOUMA, D. and McINTYRE, G.A. (1963)  A factorial
       field experiment with citrus.  *J. Hort. Sci.*,
       *38*: 175-198.

5.    CARY, P.R. (1968)  The effects of tillage, non-
       tillage and nitrogen on yield and fruit compo-
       sition of citrus.  *J. Hort. Sci.*, *43*: 299-315.

6.    —————— (1970)  Growth, yield and fruit compo-
       sition of Washington Navel orange cuttings, as
       affected by root temperature, nutrient supply
       and crop load.  *Hort. Res.*, *10* : 20-33.

7.    CHAPMAN, H.D. (1960)  Leaf and soil analysis in
       citrus orchards.  *Man. Calif. Agric. Exp. Stn.
       No. 25.*

8.    DUNCAN, D.B. (1955)  Multiple range and multiple
       F-tests.  *Biometrics*, *11*: 1-42.

9.    GROENEWEGEN, H. and BOUMA, D. (1960)  The chemical
       composition of the soil in a factorial exper-
       iment with citrus.  I: Exchangeable metal
       cations and their effect on the cation content
       of citrus leaves.  *Aust. J. Agric. Res.*, *11*:
       208-222.

10.   STEENBJERG, F. (1964)  In: *Plant Analysis and
       Fertilizer Problems IV*: 426.  Bould, C., Prevot,
       P., and Magness, J.R. (eds.).  Amer. Soc. Hort.
       Sci., New York.

*Questions to Dr. Cary*


SHOJI:  When you talk about kilograms per hectare of nitrogen for citrus orchards, how many trees per hectare do you have?

CARY:  Trees are planted at 90 trees to an acre, or 222 trees per hectare.

ALÈ:  What is the soil type?

CARY:  It is a fine sandy clay loam on the surface, overlying at a depth of about 60 cm a medium clay which contains some limestone rubble.  The pH in the virgin state varies from 6.5 to 7 near the surface, to 8.5 at 60 cm.

POULSEN:  Are you sure that when you apply phosphate on the surface it is moving down?  Do you have any culture under the citrus trees, or is it cultivated soil?

CARY:  In this particular experiment we have been comparing tillage and non-tillage systems.  However, the best treatment has been one where we have maintained a bare soil surface, controlling any weed growth by application of kerosene sprays.  This was done until 1965, and since then we have been using bromacil.

Underneath the tree itself we leave the soil undisturbed; we now have an excellent mulch composed of leaf detritus accumulated over the years.  This mulch increases the pH and overcomes to some extent the acidity obtained at high nitrogen levels.  While the virgin pH is of the order of 6.5, we can get below the tree crowns up to 7.5 at low nitrogen levels, and about 5.5 at high nitrogen levels; whereas the latter in unlimed inter-row areas reduced pH levels to as low as 4.

BOWEN:  You mentioned that you thought that total dry matter production was the thing that really mattered;

and I'd say yes, this is so with pasture.

However, with citrus I suggest that it is the fruit which is most important; similarly with timber it is the stem.

CARY:  I agree that the amount of consumable produce is a very important factor.  However, I am trying to indicate that before high yields of citrus can be obtained in any one year, it is necessary for an adequate amount of vegetative growth to be produced not only in that year, but also, and more importantly, in the preceding year.

BOWEN:  In your sand-culture experiments you showed that a high P-low N treatment caused a lower leaf nitrogen at 25°C as compared with 19°C. Can you explain this?

CARY:  It is lower because it is utilized to a greater extent.  The actual nitrogen level of the leaf may be lower because it is spread out over something like twice as many leaves.

CAMPBELL:  What percentage of the exchange capacity does your calcium represent?  Could you also give us any ideas concerning the reason for your response to the large amount of calcium?

CARY:  I can't speak with certainty on the percentages, but the exchangeable calcium, as extracted by ammonium nitrate, is of the order of 2000 ppm; that represents more than half of the exchange complex.

The response to the calcium application has not been directly, so to speak, in terms of calcium, but it has been one in which calcium has counteracted the soil acidity, which in turn has rendered the phosphorus more readily available.  In putting on calcium to the extent of producing soil neutrality, the inherent phosphorus in the soil became available and also the superphosphate that had been applied to the soil was rendered more readily available to the roots.

JUNGK: Could it be that you have a multiple deficiency which could be one of the causes for your huge differences caused by lime application?

CARY: There has been no evidence of a multiple deficiency. We have two trace element deficiencies inherent in our soils, namely zinc and manganese; these are overcome by annual applications of foliar sprays. In general the soil acidity induced by high nitrogen applications had a beneficial effect in increasing the availability of manganese. Toxicities at low pH levels also were not observed; there was never any indication of undesirably high levels of aluminium or manganese. It is possible, however, that levels of ammonia could have been undesirably high on high nitrogen plots.

# ABSENCE DE SIGNIFICATION BIOLOGIQUE PRECISE DU NIVEAU GLOBAL DES ELEMENTS MINERAUX DANS LA PLANTE ET DE LEUR FRACTION DEMEUREE SOUS LA FORME IONIQUE

## Wladimir Routchenko

*Station d'Agronomie I.N.R.A., Centre de Recherches de Bordeaux, Pont de la Maye (Gironde), France*

RESUME

La fraction des éléments métabolisables restant, à un instant et à un niveau donnés, sous la forme minérale dans les tissus végétaux correspond, compte tenu de la cinétique des réactions de synthèse, à un excédent, dû à une pénétration des ions supérieure au rythme de leur utilisation.

Leur sort est indéterminé; ils peuvent s'avérer utiles, inutiles ou nuisibles à la croissance du végétal en fonction des possibilités qu'il aura par la suite de les intégrer dans les substances organiques.

L'importance de la fraction restant sous la forme minérale dépend par conséquent à la fois de l'intensité de l'absorption et de celle du métabolisme. En l'absence d'autres critères, les indices analytiques relatifs à la concentration des éléments restant sous la forme minérale n'ont pas de signification précise, puisque leur valeur peut

Wladimir ROUTCHENKO. Maître Rech., Inst. Nat. Rech. Agron., Centre Rech. Agron., Pont de la Maye (France). né Moscou 1910; 1930 Ing. Dip. Ec. Nat. Sup. Sci. d'Agron., Lourdes; 1951 Dip. Et. Sup. Sci., Fac. Sci. Univ. Bordeaux.

être élevée aussi bien par suite d'une absorption
intense qu'en raison d'une activité métabolique
insuffisante, et diminuer pour des raisons inverses.

Pour les éléments qui ne contractent pas de
liaisons organiques permanentes, l'appréciation
portant sur la seule fraction minérale peut avoir
une signification plus précise vis-à-vis des conditions de l'alimentation de la plante en ces éléments.

Le diagnostic sur la nutrition minérale des plantes soulève de multiples problèmes et présente de nombreuses difficultés.

Ces dernières sont principalement le fait:

- du manque de parallélisme entre les variations qui
affectent la richesse alimentaire du milieu et la composition chimique de la plante;

- de l'absence de relation certaine entre la composition minérale du végétal et les rendements.

Des nécessités d'ordre pratique ont imposé l'adop-
tion d'un certain nombre de conventions concernant la
relation:

sol - composition chimique de la plante - rendement.

Le caractère hypothétique de ces conventions ne fait
de doute pour personne, mais elles servent néanmoins
de point de départ à l'expérimentation agronomique,
et de base fondamentale à la plupart des méthodes de
diagnostic portant sur la nutrition minérale des végétaux.

Au nombre de ces conventions, il convient de citer
en premier lieu celle qui admet que *l'alimentation de
la plante en éléments nutritifs dépend de la disponibilité de ceux-ci dans le sol.* Cette disponibilité ou
assimilabilité est déterminée au moyen de procédés

TABLEAU I

Indices Relatifs à la Richesse Alimentaire du Sol
et Valeurs Correspondantes de la Composition Chimique des Plantes
dans Trois Types de Sols (régime hydrique maintenu en permanence à 1,25 H.e.)

Eléments en ppm dans le sol et dans la matière fraîche de Zea Mays

| Type de sol | Azote | | Phosphore | | Potassium | | Magnésium | | Calcium | | Sodium | |
|---|---|---|---|---|---|---|---|---|---|---|---|---|
| | Sol | Plante | Sol | Plante | Sol | Plante | Sol | Plante | Sol | Plante | Sol | Plante |
| Calcaire argileux de Champagne | 1.20 | 0.334 | 0.083 | 0.112 | 0.240 | 3.666 | 0.084 | 0.167 | 6.768 | 0.905 | 0.028 | 0.069 |
| Limons argileux Touyas | 2.02 | 1.352 | 0.005 | 0.038 | 0.155 | 4.180 | 0.056 | 0.125 | 0.470 | 0.506 | 0.016 | 0.020 |
| Sable des Landes | 0.90 | 0.497 | 0.164 | 0.224 | 0.020 | 2.307 | 0.013 | 0.366 | 0.564 | 0.844 | 0.009 | 0.112 |

conventionnels, qui tiennent imparfaitement compte des multiples phénomènes physico-chimiques qui peuvent se produire au niveau du sol ou résulter de l'intervention du végétal.

Une autre convention porte sur *le rôle biologique* des éléments minéraux présents dans la plante: on suppose que, indépendamment de leur forme chimique, la totalité de ceux-ci participe à l'édification de l'organisme du végétal et détermine le rendement. Seule la notion des consommations de luxe vient tempérer ce concept. Or, nous savons que les éléments minéraux absorbés peuvent être: soit utiles, soit inutiles, soit même - dans certains cas - nuisibles au végétal suivant les possibilités qu'il aura de *les utiliser*.

Tant que les incertitudes résultant du caractère hypothétique de ces conventions ne seront pas dissipées, il sera difficile de réaliser des progrès substantiels dans le domaine du diagnostic de la nutrition des plantes.

Les exemples qui suivent illustrent le genre de difficultés auxquelles on peut se heurter et indiquent la manière dont on pourrait tenter de les résoudre.

Dans le Tableau I, à côté des indices relatifs à la richesse en éléments nutritifs de 3 sols, on a fait figurer les valeurs correspondantes de la composition élémentaire des plantes cultivées sur chacun d'eux.

Il ressort de l'examen du Tableau I que deux phénomènes distincts interviennent pour modifier la valeur du rapport entre le niveau d'un élément dans le sol et sa concentration dans la plante.

Le premier reflète la *compétition* entre le pouvoir fixateur du sol, dérivant principalement de sa constitution physique, et le pouvoir absorbant des racines. La faiblesse du pouvoir de rétention d'un

sol sableux fait que l'alimentation minérale des plantes y est plus aisée, même si le niveau des éléments
nutritifs y est plus bas que dans un sol riche en éléments fins.

Le second relève de l'intervention des *mécanismes
régulateurs* du végétal, qui font que l'absorption d'un
élément par la plante est relativement plus importante
lorsque cet élément est faiblement représenté dans le
sol que lorsqu'il y est abondant.

Les constatations de cet ordre démontrent la nécessité de perfectionner nos méthodes d'analyse du sol
et de préciser davantage la *notion de l'assimilabilité*
des éléments, mais on sait que l'analyse élémentaire
de la plante permet de dissiper l'incertitude concernant les *quantités* de différents éléments effectivement absorbés à l'exception de ceux qui ont pu être
excrétés par les racines ou lessivés à partir des
feuilles. Cependant, ces observations ne peuvent être
fournies qu'à posteriori par une analyse élémentaire
du végétal et correspondent, non pas à une situation
alimentaire bien définie, mais au résultat d'une série
de situations qui ont pu se succéder au cours de la
vie de la plante. Elles ne permettent donc pas une
étude précise de la nature des phénomènes qui ont con-
duit à un certain statut chimique de la plante et ne
démontrent pas les mécanismes de leur intervention.

Si, employée d'une certaine manière, l'analyse
élémentaire du végétal peut dissiper les incertitudes
résultant du caractère hypothétique des conventions
concernant le rapport sol - plante, elle ne permet
pas de relier la composition chimique de la plante au
rendement. Pour qu'une telle relation puisse être
mise en évidence, lorsqu'elle existe, il faudrait que
l'analyse de la plante soit à même de nous renseigner
sur la façon dont celle-ci utilise les éléments absorbés. Ainsi, on peut citer de nombreux cas où les ions
absorbés, non métabolisés, s'accumulent dans la plante
et exercent une action dépressive sur sa croissance.

Les *accidents* de végétation qui se produisent sur
de jeunes plantes alimentées en azote sous la forme
ammoniacale à un niveau excessif constituent un exemple
d'une telle situation.  Nous avons étudié ces accidents
sur maïs et tomates (W. Routchenko et J. Delmas, 1962;
W. Routchenko, 1964 et 1966; W. Routchenko et A. Aguilar, 1969) et avons démontré que les troubles observés
étaient dûs à l'*accumulation de* $NH_4^+$ à des taux toxi-
ques et non à une "acidification physiologique" du milieu, à laquelle on attribuait auparavant ces accidents.

Des troubles de l'intégration primaire des ions
métabolisables intéressent également d'autres éléments.
C'est notamment le cas du phosphore qui s'accumule sous
la forme minérale lorsqu'il y a une *carence en* Zn
(W. Routchenko, 1961).  Le processus métabolique peut
également être interrompu à un stade intermédiaire.
Nous avons observé des accumulations d'acides aminés
et d'amides et un ralentissement de la formation des
chaînes protéiques quand l'alimentation en *soufre* est
insuffisante.

Il apparaît que l'analyse de la plante ne pourrait jouer pleinement le rôle d'un instrument de diagnostic que si elle était en mesure de mettre en é-
vidence *la répartition entre les différentes formes
chimiques* des éléments présents dans le végétal à un
niveau et à un instant donnés.  Une telle analyse
pourrait conduire à l'établissement d'un jugement sur
les relations:

alimentation - composition chimique de la plante -
rendement,

si des facteurs non alimentaires du milieu ne pouvaient
modifier la signification des indices analytiques.

A titre d'exemple, nous reproduisons au Tableau II
le bulletin d'analyse se rapportant à des *maïs cultivés
sur sol de Champagne,* avec et sans *couverture plastique*

## TABLEAU II

## Bulletin D'Analyse No. 61

I. N. R. A.
Station d'Agronomie
33 - Pont de la Maye

Diagnostic de la
nutrition des plantes

Echantillons prélevés le: 10/7/69
provenant de: région parisienne
(M.H. Belin)
Espèce et variété: *Zea Mays*, INRA 258
Stade de développement: 8-9 feuilles
Organe de référence: les 2 premiers en-
trenoeuds (suc extrait par éthérolyse)

| Référence échantillons | 105 - témoin | | 107-s/plastique | |
|---|---|---|---|---|
| Indice de croissance | 28.35/4.40 | | 183.95/20.45 | |
| ELEMENTS | mg/l | méq/l | mg/l | méq/l |
| AZOTE  $NO_3^-$ | 298 | 21.27 | 314 | 22.41 |
| $NH_4^-$ | 41 | 2.92 | 50 | 3.56 |
| aminé + amidé | 639 | | 518 | |
| protéique | 188 | | 119 | |
| AZOTE SOLUBLE TOTAL | 1 166 | | 1 001 | |
| PHOSPHORE  $PO_4H_2^-$ | 7 | 0.22 | 24 | 0.77 |
| glucidique | 17 | | 40 | |
| ppté p.alcool | 70 | | 104 | |
| PHOSPHORE SOLUBLE TOTAL | 94 | | 168 | |
| SOUFRE  $SO_4^-$ | 41 | 2.55 | 46 | 2.86 |
| SOUFRE SOLUBLE TOTAL | 59 | | 98 | |
| CHLORE  $Cl^-$ | 1 453 | 40.97 | 1 135 | 32.01 |
| $\Sigma$ des anions | | 65.01 | | 58.05 |
| Potassium | 4 910 | 125.58 | 4 240 | 108.45 |
| Calcium | 175 | 8.73 | 32 | 1.59 |
| Magnésium | 208 | 10.37 | 111 | 9.12 |
| Sodium | 24 | 1.97 | 8 | 0.34 |
| $\Sigma$ des cations $NH_4^+$, $K^+$, $Ca^+$, $Mg^{++}$, $Na^+$ | | 149.57 | | 123.06 |
| $\Sigma$ cations / $\Sigma$ anions | | 2.30 | | 2.11 |

(expérimentation réalisée par l'A.G.P.M., les plantes
étant analysées par le Laboratoire expérimental de
diagnostic sur les végétaux de Pau, dont nous assurons
la couverture scientifique).

Le rendement en grain des plantes témoins n'a été
que de 9,8 q/ha alors que les plantes cultivées sous
plastique ont produit 97,1 q/ha.

L'interprétation de ces analyses serait difficile
à faire si on ne disposait que de valeurs relatives à
la composition chimique de la plante.  On pourrait seu-
lement constater que l'action d'une couverture plastique a eu pour effet d'accroître les taux des anions
métabolisables $NO_3^-$, $PO_4H_2^-$, $SO_4^{--}$ et conduisait à une
réduction de la teneur en $Cl^-$ et en cations à l'exception de $NH_4^+$.  On pourrait constater aussi une baisse
des formes organisées de l'azote et une élévation de
toutes les formes du phosphore, mais la véritable
signification de ces valeurs ne peut apparaître qu'en
se référant aux *indices de croissance* figurant sur la
première ligne du bulletin.  Il s'agit des poids de
matière fraîche et de matière sèche d'un organe de
référence, qui indique que les plantes cultivées sous
plastique ont produit environ 6 fois plus de matière
organique que les témoins. *Il en résulte que les valeurs identiques relatives à la composition chimique
correspondent en réalité, dans le cas du plastique, à
une alimentation très supérieure.*  L'effet bénéfique
du traitement apparaît principalement dans le domaine
de la nutrition *phosphorique,* l'alimentation en azote
et en soufre étant également stimulée.  L'intensifi-
cation du rythme d'organisation de ces éléments suit
la progression de leur absorption; par contre, l'alimentation en cations bivalents - surtout en calcium -
ne suit pas cette progression.

Bien que l'analyse porte sur des plantes n'ayant
que 8 feuilles, on voit que, déjà à ce stade, grâce
au recours à *l'indice de croissance* et à *une analyse*

*fournissant des renseignements tant sur les conditions
d'absorption que d'organisation des éléments,* on se
trouve en mesure de formuler un jugement sur l'action
d'un traitement.

Pour résumer ces diverses observations, on pourrait dire que *l'analyse de la plante,* employée en tant
qu'instrument de diagnostic, devrait fournir des in-
formations sur *l'absorption* et *l'organisation* des éléments nutritifs, et sur le *rythme* de croissance de la
plante étudiée.  Pour cela, il est nécessaire que
soient appréciées les *différentes formes chimiques*
sous lesquelles les éléments nutritifs se trouvent
représentés dans les tissus analysés, et qu'un système
de comparaison permette de juger le statut chimique du
végétal vis-à-vis de sa croissance.

C'est en partant de ces considérations que nous
avons mis au point la *méthode de diagnostic* employée
à la Station d'Agronomie de Bordeaux et appliquée de-
puis deux ans par un laboratoire d'analyses en série
créé à cet effet à Pau (W. Routchenko, 1967).

Les diagnostics réalisés ont apporté une large
contribution à la bonne compréhension des résultats ex-
périmentaux portant sur des sujets et des espèces végétales extrêmement variés.  Dans un certain nombre de
cas, l'emploi de la méthode préconisée a permis de résoudre des problèmes difficiles, posés par des troubles
de la nutrition des plantes dépendant de l'équilibre et
du niveau de l'alimentation minérale, comme aussi des
conditions physiques du sol et de l'environnement.

BIBLIOGRAPHIE

1.   ROUTCHENKO, W. (1961) Détermination d'une carence
     en Zn sur Maïs. *C.R. Acad. Agric.*, 11/10/61,
     739-741.

2.   ROUTCHENKO, W. et DELMAS, J. (1962) Contribution à
     l'étude des variations de la composition miné-
     rale du suc de maïs soumis à deux types d'ali-
     mentation azotée. *C.R. Acad. Sci. 254,*
     13/6/62, 4199-4201.

3.   ROUTCHENKO, W. (1964) Problèmes de la fertilisa-
     tion azotée du maïs dans' le Sud-Ouest aquitain.
     *B.T.I. 186,* 1-7.

4.   ROUTCHENKO, W. et LUBET, E. (1966) Rôle de l'équi-
     libre cationique et de chacun des principaux
     cations dans le déclenchement du phénomène
     d'intoxication ammonique de jeunes plants de
     Zea Mays. *C.R. Acad. Sci. D 262,* 10/1/66,
     281-284.

5.   ROUTCHENKO, W. (1967) Appréciation des conditions
     de la nutrition minérale des plantes basée sur
     l'analyse des sucs extraits des tissus conduc-
     teurs. *Ann. Agron. 18*:(4), 361-402.

6.   ROUTCHENKO, W. et AGUILAR, A. (1969) Contribution
     à l'étude de la nutrition minérale de la tomate
     en fonction de la forme ionique de son alimen-
     tation azotée. *Agrochimica XIII*:(4-5), 280-291.

*Questions à M. Routchenko*

PREVOT: Je considère que la communication de M. Routchenko est très importante et devrait faire l'objet d'une discussion approfondie.

COIC: Dans la compétition entre le sol et la plante, on oublie généralement un élément indispensable aux racines, qui est l'oxygène de l'air. Par exemple, lorsqu'on parle de sol sableux, qui retient peu énergiquement les ions minéraux assimilables, il faut penser que ce sol sableux est aussi un sol perméable à l'air.

ROUTCHENKO: Je crois que l'observation de M. Coïc est capitale. Dans mon exposé très bref, je ne me suis pas attardé sur certains détails, et dans les conditions physiques, bien entendu, il faut comprendre le potentiel d'oxydant du sol, et les disponibilités à l'oxygène des échanges qui peuvent se produire.

PREVEL: Je voudrais demander à M. Routchenko par rapport aux méthodes d'analyses classiques des éléments totaux, si la méthode étudiant les différentes formes des éléments dans la plante, ne subit pas de beaucoup plus fortes variations quotidiennes, entre le matin et l'après-midi, par exemple.

ROUTCHENKO: Effectivement, on assiste à des variations importantes, c'est pourquoi les conditions de prélèvement sont extrêmement rigoureuses. Nous prélevons les échantillons dans les deux heures qui suivent le lever du soleil. Ceci est une première condition et, deuxièmement, l'organe de référence prélevé est immédiatement fixé dans les terres à -40°C par l'emploi de neige carbonique. C'est-à-dire, la composition biochimique de l'organe est strictement maintenue.

PREVEL: Combien d'analyses par jour, faites-vous?

ROUTCHENKO: Deux techniciens bien entraînés peuvent analyser 24 échantillons par semaine pour ces déterminations.

# FIELD SIMULATION OF POT CONDITIONS AND GENERALIZATION. POT-DETERMINED REFERENCE VALUES IN CROP LEAVES

Pieter W.F. De Waard

*Department of Agriculture Research, Royal Tropical Institute, Amsterdam, The Netherlands*

ABSTRACT

"Feeding" a plant in physically and chemically poor soils actually entails creating a favorable nutritional environment by simulating the nutritional conditions of pot sand cultures. The necessary manipulations principally amount to ensuring root concentration, adequate soil moisture and fertilizer concentration. The proposed interventions would necessitate the use of foliar diagnosis.

Direct generalization of pot-determined reference values of plants grown under simulated pot conditions to field plants appears permissible. This hypothesis was tested on black pepper (*Piper nigrum* L.) under Sarawak conditions. Reference values for N, P, K, Ca and Mg in leaves were determined; vegetative health served as a criterion. External appearance and fruit production were used as criteria in the field trials.

Pieter W.F. DE WAARD. Sen. Agron., Soil Fertility & Crop Nutr., Dept. Agric. Res., Roy Trop. Inst., Amsterdam (Holland) since 1966. b. 1929 Medan (Indonesia); 1956 M.Sc., Wageningen; 1959 Chief, Pepper Res. Branch, Dept. Agric. Kuching (Sarawak); 1969 Ph.D., Wageningen.

In particular, diagnosis of field vines showed that, given regular fertilizer supply, satisfactory vegetative conditions and highest production tended to correspond with reference values from full-nutrient pot treatments. Similarly, lower leaf values could generally be related to lower yields and less vigorous appearance. Overall results tend to support the proposed hypothesis, indicating possible simplification of diagnosis and nutritional control in crop production.

## 1.  INTRODUCTION

Faced with the acute need to determine the nutritional status of perennial black pepper (*Piper nigrum* L.) and to recommend selective fertilizer policies, foliar diagnosis was considered to be the most appropriate method. Frequently, pot trials are employed to establish reference data for the mineral composition of leaves, but a direct generalization of these values to field conditions may not be a simple matter. Little attention is usually paid to a probably important source for this problem, associated with the fundamental difference existing between pot and field nutrition with respect to uninhibited movement of nutrients to roots. The great care involved in pot trials to offer an optimal nutrient supply to the root surface promotes maximum access to nutrients, whereas random conditions in the soilbody limit access to nutrients on account of the fact, that plant roots are forced to actively explore a large volume of soil for small irregular deposits of nutrients in severe competition with soil constituents.

This would imply that nutrient concentrations in leaves of field plants may deviate randomly from those of potted plants and therefore, pot-determined values seem of little practical value, if used for field-grown plants.

The discrepancy between pot and field conditions may be overcome by appropriate "simulation" of pot conditions in the field. The leading principle is implied in the concept of "feeding the plant" (van Diest, 1967). From work by Barber et al. (1963), Bray (1964), Jenny (1966), Lewis and Quirk (1967), Nye and Marriotth (1969) and others it may be inferred that under field conditions "feeding the plant" concurs with high nutrient accessibility, and essentially involves systematic attempts to maintain a maximum flux of the respective nutrients from soil and fertilizer to the root surface. This may be achieved by root concentration, concentrated fertilizer placement and presence of adequate water at or near field capacity.

In this paper the validity is examined of the hypothesis that pot-determined values may be directly employed as threshold values for performance of field plants, if the appropriate measures are taken for sustained access to nutrients of most active roots. Black pepper has been used as the test crop.

## 2. MATERIALS AND METHODS

### 2.1. *Pot experiment*

A single-node clonal cutting of var. Kuching was planted in a suitable 3 1-pot filled with acid-washed, coarse quartz sand with a pH of 5. After growth resumption, pots were transferred to a green house bench, where conditions were suitable for proper growth and development. The sand surface was covered with finely perforated, black polyethylene discs to prevent growth of algae and to limit evaporation, while adequate ventilation was ensured.

A somewhat modified Hewitt solution was adopted for

complete treatment (de Waard, 1969). Other treatments
included: omission of N, P, K, Ca and Mg in turn. The
pH of the dilute nutrient solution was adjusted to 5.0.
Plants received 0.5 l solution per pot every 2 days. A
non-limiting moisture regime existed during the experiment. There were six replicates in all.

The experiment consisted of 2 major stages coupled
by a transitional stage, as follows:

*Stage 1.* For each deficiency treatment pots received 25% of the full concentration present in the complete solution.

*Stage 2.* A gradual transition was introduced towards full deficiency by giving nutrients at 10% of the
full level for 1 week. No leaves were collected.

*Stage 3.* Complete deficiency conditions were established. Characteristic symptoms of deficiency soon
developed.

In stages 1 and 3 leaves were sampled, cleaned,
prepared and dried in accordance with a carefully tested
standard procedure (de Waard, 1969). Chemical analysis
was carried out according to methods described by van
Schouwenburg (1966).

## 2.2. *Field experiments*

Six groups of 10 berry-producing vines of uniform
age and external healthy appearance were selected in
January. Annual yields ranged from 4 - 15 kg fresh
berries/vine (series 1). Mild symptoms of N, K and Mg
deficiency were present in 6 groups of 10 vines, selected
next April; crop yields varied from 5 - 16 kg berries/
vine (series 2). In July next, 3 groups of 20 vines
were selected. These groups showed N deficiency but

there was no die-back of branches (series 3).

A blanket dressing of 4.4 kg dolomite/vine was broadcast on the mound, and lightly worked in. Compound fertilizer containing 12% N, 12% $P_2O_5$, 17% $K_2O$ and 2% MgO was applied at 2.7 kg/vine to 5 groups of series 1 and all groups of series 3; the one group of series 1 and all groups of series 2 received NPK fertilizer without Mg. In September (early in the rainy season) the first of 4 split dressings was given and placed in 2 narrow bands, some 10 cm below the soil surface, on either side of the underground stem, in close proximity to the feeder roots. This was repeated at monthly intervals until January. Following fertilizer application flower development proceeded until January; from this time onwards until the harvest in July berry development is the predominant physiological activity.

A cemented layer of subsoil was present at a depth of 90 cm, serving as the "bottom" of a gigantic soil pot. Very little root penetration was observed below this depth (de Waard, unpublished). Rainfall was adequate and well distributed to maintain a pF value between 2 and 2.8 throughout the year, at a depth of 10 cm. Air temperature and humidity show steady diurnal cycles.

Standardized leaf sampling was carried out in January, April, July and January for each of the respective series. Pretreatment and chemical analysis proceeded as for the pot trial. Calculations were carried out with the aid of an IBM 1630 computer.

## 3. RESULTS

### 3.1. Pot experiment

The results of foliar analysis are presented in Table 1. On complete solution leaf N, P, Ca and Mg are closely similar in stages 1 and 3. Leaf K in stage 3 appears somewhat below that in stage 1. For comparison, controls with complete nutrient solution were carried along throughout the 3 stages.

Scrutiny of the data shows that deficiencies of the individual elements in solution concur with highly significant lower levels in the leaves for the respective elements in the intermediate and deficient stages as compared with those in the complete solution. Very clearcut antagonistic mechanisms appear to exist for particular combinations of elements. These may be subdivided into 3 groups: (1) N vs. P, when N is deficient, (2) K vs. Ca and Mg, when K is deficient, and (3) Ca vs. Mg and vice versa, when one element is deficient; omission of P does not induce any antagonistic mechanism. Based upon this information normal, critical and deficient reference values have been tentatively established (Table 2).

From these apparently regular patterns it may be inferred that in the presence of sufficient nutrients, mass flow and diffusion are able to supply adequately all nutrients but the deficient one under the given pot conditions. If physical limitations should hamper mass flow and diffusion to the root surface this effect is likely to be conducive to patterns of complicated nutrient shortages, irrespective of nutrient levels in the medium. Consequently, validity of pot-determined reference values in particular may not be consistent from year to year and from place to place.

TABLE 1. The Effect of Intermediate and Complete Deficiency of a Single Element on the Nutrient Composition in Leaves of Black Pepper.

| Treatments | Nitrogen | | Phosphorus | | Potassium | | Calcium | | Magnesium | |
|---|---|---|---|---|---|---|---|---|---|---|
| | ID | CD | ID | CD | ID | CD | ID | CD | ID | CD |
| | % dr.m. | % dr.m. | % dr.m. | % dr.m. | % dr.m. | % dr.m. | % dr.m. | % dr.m. | % dr.m. | % dr.m. |
| Full | 3.41 | 3.10 | 0.18 | 0.16 | 4.31 | 3.38 | 1.68 | 1.66 | 0.44 | 0.45 |
| Nitrogen omitted | 2.99** | 2.32** | 0.22** | 0.23** | 4.68 | 3.65 | 1.56 | 1.37 | 0.33* | 0.38 |
| Phosphorus omitted | 3.44 | 3.35 | 0.11** | 0.09** | 4.06 | 3.52 | 1.66 | 1.66 | 0.42 | 0.47 |
| Potassium omitted | 3.55 | 3.14 | 0.18 | 0.17 | 2.62** | 1.99** | 2.29** | 1.87 | 0.56** | 0.63** |
| Calcium omitted | 3.53 | 3.31 | 0.18 | 0.17 | 4.16 | 3.46 | 1.12** | 0.86** | 0.69** | 0.73** |
| Magnesium omitted | 3.52 | 3.34 | 0.17 | 0.18 | 4.50 | 3.98 | 1.97** | 2.22** | 0.14** | 0.18** |
| Standard deviation | 0.21 | 0.27 | 0.023 | 0.025 | 0.55 | 0.64 | 0.16 | 0.25 | 0.068 | 0.070 |
| Coefficient of variation | 6.2% | 8.8% | 13.0% | 15.0% | 13.6% | 19.3% | 9.3% | 15.7% | 15.8% | 14.8% |

\* Significant at $P = 0.05$ } with respect to
\*\* Significant at $P = 0.01$ } "full nutrients"

ID = Intermediate level of deficiency
CD = Complete deficiency

TABLE 2. Reference Values for N, P, K, Ca, Mg and log
N/P, Proposed with Respect to the Vegetative Condition
of Black Pepper Vines; these Values have been Determined in 3-liter Pots Using a Sand Medium

| | Classification | | | |
|---|---|---|---|---|
| Elements | Normal | Critical | Deficient | Indicating |
| N (% dr.m.) | 3.40-3.10 | 2.80-2.70 | <2.70 | N deficiency |
| log N/P | 1.28 | 1.22-1.14 | <1.13 | N deficiency |
| P (% dr.m.) | 0.18-0.16 | 0.14-0.10 | <0.10 | P deficiency |
| K (% dr.m.) | 4.30-3.40 | 2.62-2.00 | <2.00 | K deficiency |
| Ca(% dr.m.) | 1.68-1.66 | 1.20-1.00 | <1.00 | Ca deficiency |
| Mg(% dr.m.) | 0.45-0.44 | 0.30-0.20 | <0.20 | Mg deficiency |

## 3.2. Field experiments

In Figs. 1 and 2 the effect of leaf N and P with
yield is plotted for the series 1, 2 and 3 in the 1st
and 2nd year of observation. The relations show that
normal levels (Table 2) of these elements in January,
at the end of flowering, are bringing about a high production in that year; moreover, normal levels are
necessary to maintain a healthy and vigorous vegetative
condition.

The close correlation between leaf N and P, as
found in the pot trial, may be clearly seen by intro-
ducing the ratio N/P; for clarity log N/P values have
been used. Plotting shows that yields may be high at
relatively high values and *vice versa*. After a sharp
decrease from January to July, owing to rising leaf P
and decreasing leaf N, yields in the next year are low.
The relation of log N/P values and yield is fairly consistent, irrespective of series and year.

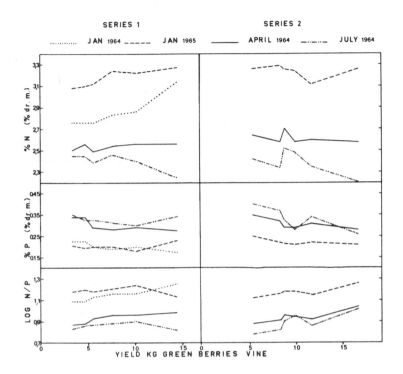

FIG. 1.   Leaf N, leaf P, log N/P in leaves and
          yield over the period Jan. 1964-Jan.
          1965.  High yields tend to coincide
          with the proposed normal values for
          N and P in plants with a healthy veg-
          etative condition (% dr.m.)

The combined results imply that manipulation of
leaf N and P aimed at maintaining proposed normal
levels in the field at appropriate times is essential
for good crop production and may well be achieved in
practice.

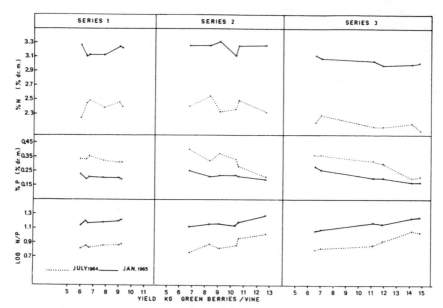

FIG. 2.   Leaf N, leaf P, log N/P in leaves
          and yield over the period July 1964–
          July 1965.  Note overall similarity
          with data in Fig. 1 (% dr.m.)

Yields and leaf K, Ca and Mg are presented in
Figs. 3 and 4.   In the range of some 8 - 14 kg
berries/vine overall results seem to imply that high
crop production at levels below 2% for K, 1.7% for Ca,
and 0.20% for Mg (Table 2) entails serious vegetative
deterioration and severe damage to the production
potentials of the vines, as the canopy displays
serious symptoms of K and Mg deficiency accompanied by
severe die-back of branches.

In contrast, in the range of 4 - 8 kg berries/
vine, under the same environmental conditions, there
is a positive relation with leaf K in January and in
July; over this period leaf K reduction is negligible
although a concurrent rise of leaf Ca may still be
observed.   This was also found in the pot trial.   Leaf
Mg tends to decrease, but remains within the fair range

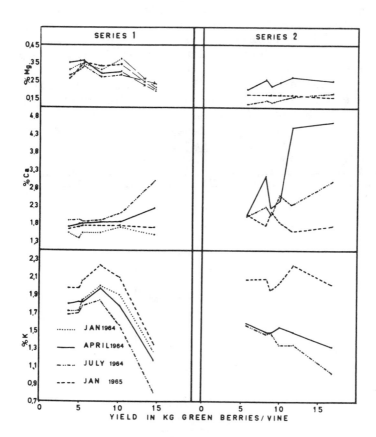

FIG. 3.   K, Ca and Mg in leaves and yield over
          the period Jan. 1964-Jan. 1965.  Extra-
          polation of that part of the Jan. 1964
          curve between 4 and 8 kg/vine indicates
          that 25 kg/vine may be borne on vines
          following increase of leaf K to the
          proposed normal value of some 3.3-3.4%
          in January.  No subsequent deficiency
          of K and detrimental plant exhaustion
          is observed (% dr.m.)

(Table 2).  Symptoms of deficiency and vine deterioration were absent.  Overall results indicate that
relatively low yields may coincide with fair to
critical leaf K, whereas in the case of high yields,
high levels of K are needed to prevent irreparable
physiological exhaustion.  The data seem to imply that
each yield size requires a specific "minimum" value
for leaf K, below which chances of physiological
exhaustion exist.  These values concur with the proposed "normal" values for Ca and Mg in January.

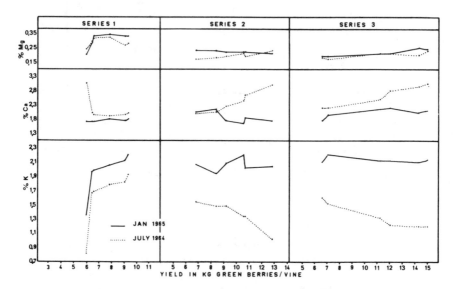

FIG. 4.   K, Ca and Mg in leaves and yield for
          the period July 1964–July 1965.  Gen-
          erally, extrapolation in the series 1
          and 2 gives similar results as in Fig.
          3.  In series 3, leaf K seems in ex-
          cess for yields between 6 and 10 kg/
          vine, whereas there appears to be a
          short supply for yields over 10 kg/
          vine (% dr.m.)

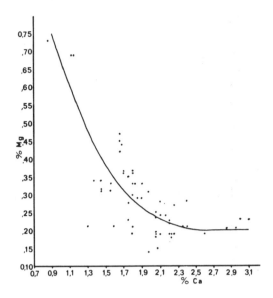

FIG. 5.  Antagonistic relations of Ca and Mg
         in leaves.  The normal value for Ca
         corresponds fairly well to that of
         Mg (% dr.m.)

Extrapolation of the 4 - 8 kg section of the leaf
K, Ca and Mg curves for January (Fig. 3) suggests that
3.3-3.4% leaf K may be adequate to support a crop of
some 25 kg berries/vine.  In this case leaf K diminishes to some 2.4% in July, remaining above the deficiency value.  Leaf Ca and Mg coincide consistently
with normal values.  Apparently, proposed normal values
for leaf K may coincide with high and low yields without the sever repercussions on vegetative condition,
whereas high yields and below normal values may entail
severe vine deterioration and complex 3-way base interactions (Fig. 4, series 2 and 3).

In Figs. 5, 6 and 7 the Ca-K, Ca-Mg, and K-Mg
inter-relations are plotted.  The antagonistic mechanisms, observed in the pot trial are manifest below

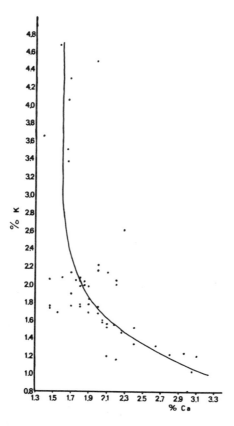

FIG. 6.    Relations of K and Ca in leaves.  If
           leaf K decreases towards the proposed
           threshold value of 2%, the proposed
           normal value for Ca tends to be main-
           tained; below 2% leaf Ca behaves
           antagonistically to leaf K (% dr.m.)

the proposed deficiency value for leaf K.  Ca-Mg
correlation shows a similar picture in that the proposed normal value of Ca, projected on the curve tends
to coincide with that of Mg.  Also the antagonistic
behavior observed in pots is evident in the field.
Antagonistic and synergistic mechanisms are present for
K-Mg (Fig. 7).  The abrupt change in the slope of the
line occurs at a value of some 2% for leaf K.

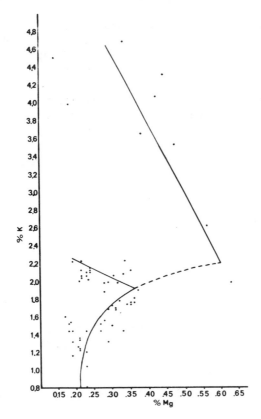

FIG. 7.   The relations of leaf K and leaf Mg.
          Note, that the intercepts occur at
          the proposed threshold value of 2%
          for leaf K (% dr.m.)

The correlation of Mg and K explains satisfactorily
that leaf K below some 2% may coincide with rising leaf
Ca and falling Mg in place of the expected Mg rise.  It
may also be observed that below 2% for K the complex
inter-relationships manifest themselves.

Generally speaking, the combined evidence supports the proposed hypothesis; the tentative reference values which have been obtained in a pot trial apparently possess a satisfactory predictive value with respect to both yield size and vegetative condition in the field. These pot determined reference values may thus be accepted as a basis for interpretation to predict fertilizer requirements of pepper, bearing in mind that appropriate measures to "feed the plant" are introduced *a priori* by "simulating" pot conditions in the field soil.

## 5.  DISCUSSION

Nutrient environment and root uptake as interfering factors influencing interpretation of leaf analytical results have been discussed before (Emmert, 1961). Bould (1968) formulated that in the absence of non-nutritional limiting factors leaf analysis provides an accurate tool for nutritional diagnosis. In this current paper attention is paid to the vital role of uninhibited nutrient transport to the root surface to allow consistent interpretation of leaf analytical results in time and place.

Limitation of nutrient flux should necessarily influence the rate of uptake and it may be considered responsible to a large proportion for the generally wide range usually quoted for accepted optimal values of perennial crops (Chapman, 1966). For oil palm Ferwerda (1961) reported differences in leaf concentrations with respect to area and year, as compared with the normal reference values. A considerable influence of environment on leaf composition was found in olive (Bouat, 1956); for one variety leaf concentrations varied with respect to defferences in location. Lévy (1956) suggested introduction of an "economic optimum" and an "experimental optimum."

Yamasaki and Kawamura (1967) reported recently that uptake of N, P, K, Ca and Mg by Satsuma orange in pots with different soil types was maximum at a pF value of 2-2.8 in the presence of an adequate flux of nutrients; leaf concentrations were also maximum at pF values near or at field capacity. The apparently ready uptake concurred with optimal growth performance. The concentrations of an individual element in comparable leaves, were in general of a similar magnitude, irrespective of soil type.

Transport phenomena of nutrients in the soil have been extensively investigated by, amongst others, Barber et al. (1963); Fried and Broeshart reviewed the subject comprehensively (1967). Nye and Marriotth (1969) recently reported on computer simulation of transport processes in the soil to the root surface. Water transport processes have been reported on, amongst others, by Bolt (1966) and Groeneveld (1969). The consensus of opinion amounts to (1) high concentrations of nutrients in soil solution are essential for an uninhibited flux of nutrients to the root surface by mass flow and diffusion, (2) higher moisture tensions considerably reduce the mass flow and diffusion, and the nutrient flux to the root surface, (3) diffusion occurs over short distances only and becomes relatively more important when mass flow is reduced, and (4) physical barriers owing to soil compaction or poor structure adversely affect mass flow and diffusion.

The current work elaborates on this theme. An adequate nutrient flux was expected to occur by simulating pot conditions in the field. Data on nutrient concentrations in leaves, grown under conditions of the principle of "feeding the plant", yield and vine performance were in good agreement with relevant reference values in the pots. Thus it appears justified to compare field values directly to reference values from pots to predict nutritional conditions for black pepper, if prevailing nutrient transport conditions in the medium

ensure uninhibited ion mobility from soil and applied
fertilizer to reach the soil/root interface.

The theoretical considerations outlined above in
combination with past experience and recent developments
seem to point to the view, that in leaf analysis so far
most attention has been paid to stratification and stan-
dardization within the above-ground parts of the plants.
In contrast, the influence of the mechanics of nutrient
supply to the underground parts has been somewhat neg-
lected. By reversing the approach, determination of
reference values and the interpretation of foliar diagnosis may become more simplified.

REFERENCES

1.   BARBER, S.A., WALTER, J. and VASEY, E.H. (1963)
       Mechanisms for the movement of nutrients from
       the soil and fertilizer to the plant root.
       *J. Agric. Soil Chem. 11*: 205-7.

2.   BOLT, G.J. (1966)  Soil physical processes as bound-
       ary conditions to ion uptake. *Techn. Rep. Series
       Int. Atomic En. Agency No. 65.*

3.   BOUAT, A. (1956)  La fumure de l'olivier. La sol-
       ution apportée par le diagnostic foliaire.  In:
       Wallace, T. (ed.), *Plant Anal. Fertilizer Probl.,*
       311-22. I.R.H.O., Paris.

4.   BOULD, C. (1968)  Leaf analysis as a diagnostic
       method and advisory aid in crop nutrition. *Exp.
       Agric. 4*: 17-27.

5.   BRAY, R.H.A. (1964)  A nutrient mobility concept
       of soil-plant relationships. *Soil Sci. 78*: 9-22.

6.  CHAPMAN, H.D. (1966) *Diagnostic Criteria for Plants and Soils*. Univ. of California, Div. of Agric. Sciences.

7.  DIEST, A. VAN (1967) *Bemest Men de Plant of de Grond?* 20 pp. H. Veenman en Zonen, Wageningen.

8.  EMMERT, F.H. (1961) The bearing of ion interactions on tissue analysis results. In: Reuther, W. (ed.), *Plant Anal. Fertilizer Probl.*, 231-44. American Inst. of Biological Science, Washington-6, D.C.

9.  FERWERDA, J.D. (1961) Growth, production and leaf composition of the African oil palm as affected by nutritional deficiencies. In: Reuther, W. (ed.), *Plant Anal. Fertilizer Probl.*, 148-58. American Inst. of Biological Science, Washington-6, D.C.

10. FRIED, M. and BROESHART, H. (1967) *The Soil-Plant System in Relation to Inorganic Nutrition*. Monograph. Academic Press, New York and London.

11. GROENEVELT, P.H. (1969) *Koppelingsverschijnselen bij Transport Processen in de Bodem*. Ph.D. thesis. Agric. University, Wageningen.

12. JENNY, H. (1966) Pathways of ions from soil into roots according to diffusion models. *Plant Soil* 25: 265-89.

13. LEWIS, D.G. and QUIRK, J.P. (1967) Phosphate diffusion in soils and uptake by plants, iii $P_{31}$ movement and uptake as indicated by $P_{32}$ autoradiography. *Plant Soil 26*: 445-53.

14. LÉVY, J.F. (1956) Resultats obtenus grâce au diagnostic foliaire. In: Wallace, T. (ed.), *Plant Anal. Fertilizer Probl.*, 384-94. I.R.H.O., Paris.

15.  NYE, P.H., and MARRIOTTH, F.H.S. (1969)  A theor-
     etical study of the distribution of substances
     around roots resulting from simultaneous dif-
     fusion and mass-flow.  *Plant Soil 30*: 459-72.

16.  SCHOUWENBURG, J.C. VAN (1966)  *Voorschriften voor
     Gewas-Analyse.*  Agric. University, Wageningen,
     Holland.

17.  WAARD, P.W.F. DE (1969)  *Foliar Diagnosis, Nu-
     trition and Yield Stability of Black Pepper
     (Piper nigrum L.) in Sarawak.*  Communication
     No. 58, Royal Tropical Institute, Amsterdam,
     Holland.

18.  WAARD, P.W.F. DE  Unpublished results.

19.  YAMASAKI, L. and KAWAMURA, A. (1967)  Studies on
     the growth of citrus trees in relation to the
     soil-water system.  I: Effects of the soil
     moisture on the growth of young Satsuma orange.
     *Bull. Shikoku Agric. Exp. Stat., Japan 17*:
     13-46.

*Questions to Dr. de Waard*

SHEAR:  Why do you have to have so many parts to get a balanced figure?  Why wouldn't a ratio between these elements do just as well?

DE WAARD:  The ratio is not enough.  We have to know how much of each nutrient lies in the plant.

SHEAR:  Does this then hold regardless of what your levels of the other elements would be; or do you have another set when you change the levels of the others?

DE WAARD:  If we try to find out what will happen when we deal with different amounts of potassium, calcium, and so on, the whole system will grow.  We have to put that into a system of 7 to 9 figures.

ROUTCHENKO:  Ne croyez-vous pas qu'il est difficile d'établir par voie expérimentale l'équilibre N:K, sans parler de la forme ionique de l'azote? et que si vous apportez votre azote sous forme ammoniacale, vous allez deprimer l'absorption du potassium; si vous l'apporterez sous forme $NO_3$, vous allez favoriser l'absorption du potassium.  Ne croyez-vous pas qu'il est indispensable de préciser, dans une expérience de ce genre, la forme sous laquelle l'action du sol fourni est absorbé par les deux étapes.

DE WAARD:  If you use different kinds of fertilizer, for instance - sulphate of ammonia and sodium nitrate - we will have different figures.  But in a way that is a good thing, because it enables us to control the plant concentration for the maximum growth and discover a special nutrient composition in the plants.  We don't play those different forms of nutrients against each other, and are able to use just those nutrients which give us the quality in the mature plant which we want to have.  We are able to get a good crop, a large crop;  and we are able to get a crop with the quality which we want.

# DIAGNOSIS AND CONTROL OF NUTRITIONAL DISORDERS IN CEREALS BASED ON INORGANIC TISSUE ANALYSIS

## J. Møller Nielsen

*Department of Soil Fertility and Plant Nutrition,
Royal Veterinary and Agricultural University,
Copenhagen, Denmark*

ABSTRACT

This paper deals with results of chemical plant analyses from pot experiments and field trials in various parts of Scandinavia, with a wide range of applications of various nutrients to different soils and cereals. These results form a basis for evaluating nutrient status of plants at various stages of development, and for formulating relationships between yield and nutrient concentration and between the concentrations of various nutrients at different nutrient levels.

The N, P, K, Na, Ca and Mg content of young plants can be used for diagnostic purposes--to evaluate the nutrient status, and for therapeutic purposes--to adjust, by choosing type and amount of fertilizer, the chemical composition of the plants so as to ensure a final yield of desired size and quality. Chemical analyses of the final crop may be used to elucidate both nutrient conditions in the soil during the growth period and the quality of the crop.

The correlation between the prognoses--based on chemical analyses of young plants--on the size and chemical composition of the final yield and the experimentally obtained results shows the efficacy of the proposed diagnostic method.

Jens Møller NIELSEN. Lecturer Soil Fertility & Plant Nutr., Roy. Vet. & Agric. Univ., Copenhagen (Denmark) since 1967. b. 1928 Copenhagen; 1953 Cand. Agric. and 1957 Lic. Agr., Roy. Vet. & Agric. Univ., Copenhagen.

Pot-culture and field experiments were conducted in our department from 1961 to 1969 in order to develop a quantitative diagnosis and control method for determining and adjusting the nutrient status of spring-sown cereals on the basis of the chemical composition of the plants.

## OBJECTIVE

The formulation of relationships between the chemical composition of the plants at various stages of development, and the nutrient conditions in the soil, as well as the yield and chemical composition of the plants at maturity.

## EXPERIMENTAL BASIS

Chemical analyses for a number of plant nutrients were made in different spring-sown cereals at different times in the period of growth under varying climatic, soil, and fertilization conditions. On this basis, the factors affecting the determination of the relationships mentioned under "Objective" are eliminated, or corrections made. Such factors are: climate, soil, stage of development, and species of plants. In addition to these factors, the Steenbjerg effect, the mutual effects of the nutrients and their translocation, influence the relationships mentioned. The eliminations and corrections were made on the basis of experimental and theoretical investigations, the results of which will be published elsewhere.

On the basis of the experimental results available, a method of diagnosis, prognosis and therapy has been developed.

DIAGNOSIS, PROGNOSIS AND THERAPY MODEL

Figs. 1 - 6 represent a simplified model for diagnosis and therapy for barley. The starting point is the concentration of the nutrients in the dry matter of the above-ground parts of the plants at an early stage of development.

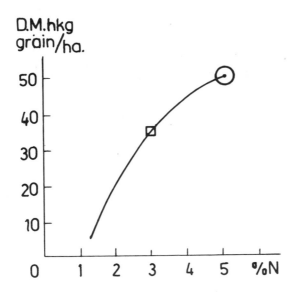

FIG. 1.  Production of dry matter (D.M.) in grain as a function of the N concentration at different degrees of pure N deficiency. Pure N deficiency is defined as an N deficiency unaffected by deficiency or excess of other nutrients.

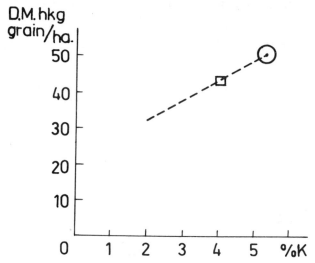

FIG. 2. Production of dry matter in grain as a function of the K concentration at different degrees of pure K deficiency.

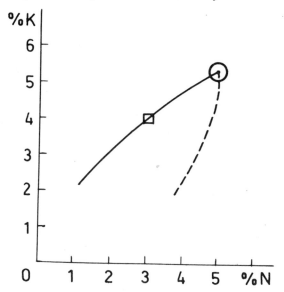

FIG. 3. K concentration as a function of the N concentration: ——— at pure N deficiency, --- at pure K deficiency.

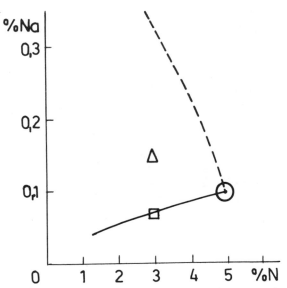

FIG. 4.   Na concentration as a function of the N concentration at pure N and K deficiency.

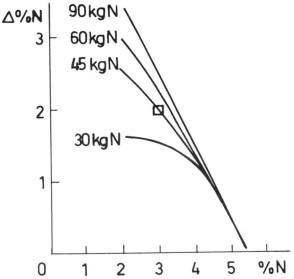

FIG. 5.   Increase in the N concentration (Δ%N) as a function of the N concentration at applications of 30, 45, 60 and 90 kg N/ha.

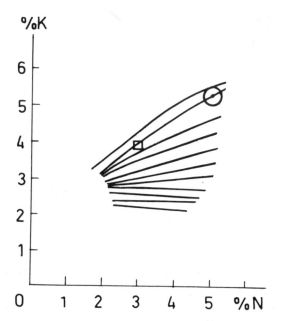

FIG. 6.   K concentration as a function of increasing N
concentrations caused by application of N.


Figs. 1 and 2 are used to determine absolute concentration deficiency, dominant nutrient deficiency,
and yield prognoses; Figs. 3 and 4 to determine relative concentration deficiency; Fig. 5 to determine N
therapy, and Fig. 6 to determine the effect of the N
therapy on the K concentration.


EXAMPLES OF THE APPLICATION OF THE METHOD


*Example 1:*   5.0 % N, 5.3 % K, and 0.10 % Na (marked ⊙
in the figures).

Fig. 1.  The state of N nutrition (5 % N) ∿ maximum
yield = *optimum N nutrition.*

Fig. 2.  The state of K nutrition (5.3 % K) $\sim$ maximum yield = *optimum K nutrition*.

*Diagnosis:*   *Optimum N and K nutrition.*

*Prognosis:*   Maximum production of dry matter in grain = *50 hkg/ha.*

*Therapy:*     *None.*

*Example II:* 3.0 % N, 4.0 % K, and 0.07 % Na (marked ⊡ in the figures).

|  |  | Concentration | | Absolute concentration deficiency |
|---|---|---|---|---|
|  |  | Optimum | Actual |  |
| Fig. 1 | State of N nutrition: | 5.0 % N – 3.0 % N = | | 2.0 % N |
| Fig. 2 | State of K nutrition: | 5.3 % K – 4.0 % K = | | 1.3 % K |

Fig. 1.  N deficiency reduces the yield of grain to 35 hkg/ha;

Fig. 2.  K deficiency reduces the yield of grain to 45 hkg/ha, i.e. *the N deficiency is dominant.*

The state of K nutrition is now investigated more closely in relation to the dominant N deficiency at 3.0 % N in Fig. 3.  The curve showing the pure N deficiency is the solid one.

| | | Concentration | | Relative K concentration deficiency |
|---|---|---|---|---|
| | | Relative optimum | Actual | |
| Fig. 3 | State of K nutrition: | 4.0 % K | - 4.0 % K = | 0 % K |

i.e. *no relative K deficiency or excess* giving further reduction of the yield already reduced by N deficiency.

Fig. 4.  The Na concentration (0.07 % Na) as a function of the N concentration (3.0 % N) supports the information previously obtained: *pure N deficiency*.

*Diagnosis:*  N deficiency.  *There is a 2.0 % N lack. Optimum K nutrition* (4.0 % K) *according to* Fig. 3.

*Prognosis:  35 hkg dry matter in grain/ha.*

An estimate is now made of the amount of N necessary to increase the N concentration from 3.0 % N to an optimum of 5.0 % N, i.e. an increase by 2 % N.

Fig. 5.  An increase of the N concentration by 2 % N ($\Delta$ 2.0 % N) from 3 % N will be obtained by an application of 45 kg N/ha.

*Therapy :   45 kg N/ha.*

Then it is investigated how such N therapy will affect the concentration of other nutrients in the plant - in this paper shown for potassium only, Fig. 6.

Fig. 6.  The application of 45 kg N/ha causing an increase of the N concentration from 3.0 % N to 5.0 % N brings about an increase of the K concentration

from 4.0 % K to 5.3 % K.    Thereby an optimum composition of N and K in the plant is obtained, i.e.
5.0 % N and 5.3 % K.

*Prognosis for the chemical composition* of the young
plant:   *5.0 % N and 5.3 % K.*

*Prognosis for the yield* at maturity:   *50 hkg dry
matter in grain/ha* (see Figs. 1 and 2).

*Example III:*   3.0 % N, 3.0 % K, and 0.15 % Na.

|  |  | Concentration | | Absolute concentration deficiency |
|--|--|--|--|--|
|  |  | Optimum | Actual |  |
| Fig. 1. | State of N nutrition: | 5.0 % N | − 3.0 % N = | 2.0 % N |
| Fig. 2. | State of K nutrition: | 5.3 % K | − 3.0 % K = | 2.3 % K |

Fig. 1.   The N deficiency reduces the grain yield to
         35 hkg/ha.

Fig. 2.   The K deficiency reduces the grain yield to
         38 hkg/ha, *i.e. the N deficiency is dominant.*

      Now the state of K nutrition is investigated in
relation to the dominant N deficiency at 3.0 % N in
Fig. 3.

| | | Concentration | | Relative K concentration deficiency |
|---|---|---|---|---|
| | | Relative optimum | Actual | |
| Fig. 3 | State of K nutrition: | 4.0 % K | − 3.0 % K = | 1.0 % K |

i.e. a *K deficiency of 1.0 % K,* further reducing the yield already reduced by the N deficiency.

Fig. 4. The Na concentration (0.15 % Na) as a function of the N concentration (3.0 % N) supports the information previously obtained: *N and K deficiency,* (see the point in Fig. 4 marked by Δ).

*Diagnosis:*  N deficiency. *A deficiency of 2.0 % N.*
                K deficiency. *A deficiency of 1.0 % K.*

It is now investigated how much the K deficiency further reduces the yield already reduced by the N deficiency.

Fig. 2. A K concentration deficiency of 1.0 % K may reduce the yield from 50 to 46 hkg dry matter in grain/ha. Such yield reduction of 4 hkg dry matter in grain/ha corresponds to a yield reduction of 8 %. The yield reduced to 35 hkg dry matter of grain/ha by the N deficiency is thus further reduced by the K deficiency, presumably by 8 %.

*Prognosis:*  35 hkg dry matter of grain/ha less 8 % =
                35 − 3 = *32 hkg dry matter of grain/ha.*

Finally it is to be investigated what amount of K must be applied to raise the K concentration from 3.0 % K to 4.0 % K and what effect such increase in the

K concentration will have on the N concentration.  No
figures have been shown here to illustrate this.
Figures in a more complete model might, for instance,
show that an application of 100 kg K/ha would raise
the K concentration from 3.0 % K to 4.0 % K and that
this increase in the K concentration would not affect
the N concentration.  The conclusion is:

*Therapy : 45 kg N/ha and 100 kg K/ha.*

*Prognosis for the chemical composition* in the young
plant: 5.0 % N and 5.3 % K.

*Prognosis for the yield* at maturity: *50 hkg dry
matter of grain/ha.*

A more complete method for quantitative diagnosis,
prognosis, and therapy, not given here, also comprises
the plant nutrients: P, Mg, and Ca determined at
various stages of development and other cereals than
barley.  Furthermore, the method comprises the composition of nutrients in the mature crop, allowing conclusions to be drawn from the young plant to the
mature plant and vice versa regarding all the
relationships mentioned under "Objective".  In other
words it is possible to control the chemical composition of the young plants with a view to both size
and chemical composition of the mature crop, or to
draw conclusions from size and chemical composition
of the mature crop to the nutrient status of the young
plants as a basis for rational fertilization of the
following crops.

The usefulness of the method is affirmed by the
close correlation found between prognoses and actual
results.

# VALUE OF VARIOUS TISSUE ANALYSES IN DETERMINING THE CALCIUM STATUS OF THE APPLE TREE AND FRUIT

Cornelius B. Shear and Miklos Faust

*Crops Research Division, Agricultural Research Service, U.S. Department of Agriculture, Beltsville, Maryland, U.S.A.*

## ABSTRACT

Since low calcium is associated with many fruit disorders, we studied its uptake, translocation, and accumulation in apple tissues by means of $^{45}$Ca movement in seedlings and by analyses of leaves and fruits from trees growing in controlled nutrient cultures. Ca moves, probably largely by exchange on lignin, to areas of high metabolic activity, in growing tissues, and accumulates in the vascular system of older tissues. Ammonium, as contrasted with nitrate nutrition, increases the translocation to and accumulation of Ca in the younger leaves. Spraying with solutions of ammonium salts has a similar effect. Increased boron in the nutrient supply or from foliar sprays increases Ca accumulation, especially at marginal levels of Ca supply. Leaf and fruit Ca does not increase in proportion to increases in Ca supply. A supply of Ca sufficient to prevent leaf symptoms of Ca deficiency is not sufficient to prevent fruit symptoms, especially in large, rapidly growing fruit. The percentage of Ca in the leaves increases with age, and decreases with age in the fruit. Thus, the later the sampling date, the poorer the correlation between leaf

Cornelius B. SHEAR. Plant Physiol. (Min. Nutr.), Agric. Res. Serv., U.S. Dept. Agric., Beltsville, Maryland (U.S.A.) since 1964. b. 1912 Vienna (Virginia, U.S.A.); 1934 B.Sc. and 1938 M.Sc., Univ. Maryland. 1935 Plant Physiol. (Min. Nutr. Research), U.S. Dept. Agric. at Beltsville (Md.) and 1939 at Gainesville (Fla.).

and fruit Ca.  Ca concentration varies greatly within
the apple fruit; therefore, selection of tissue for
analysis should be based on the part of the fruit
involved in the disorder under study.

─────────

Calcium deficiency symptoms or yield response to
Ca fertilization occur only rarely under orchard conditions.  The association of low levels of fruit Ca
with increased incidence of corking (Faust and Shear,
1968) and other storage disorders of apples (Perring,
1968), however, has emphasized the importance of Ca in
apple nutrition.

The relative immobility of Ca after deposition in
plant tissues is well documented (Wiersum, 1966).

To make effective use of tissue analyses in diagnosing and altering the Ca status of the various parts
of the apple tree, we need specific knowledge of the
movement of Ca in the tree and of the factors that
influence Ca movement to, and accumulation in, various
tissues.  We have studied Ca uptake, translocation,
and accumulation in apple tissues by means of $^{45}$Ca
movement in seedlings and by analyses of leaves and
fruits from trees growing in controlled-nutrient cultures, soil cultures, and from orchard trees.

MATERIALS AND METHODS

For the $^{45}$Ca studies, seedlings from open-polli-
nated seeds of 'York Imperial' apples were grown for 6
to 8 weeks in quartz sand or in aerated solution cultures receiving complete nutrients, and were transferred to 1-liter brown plastic bottles for
differential Ca treatments (Table 1).  One week after

TABLE 1

Nutrient Solution Used to Grow Seedlings
for Studies with $^{45}$Ca

| Nutrient | meq/l | Nutrient | ppm |
|----------|-------|----------|-----|
| N | 12 | B | .50 |
| K | 4 | Mn | .46 |
| Ca | 1, 8, 16 | Fe | 5.00 |
| Mg | 4 | Zn | .50 |
| $HPO_4$ | 2 | Cu | .25 |
| $SO_4$ | 5.5 | Mo | .50 |
| Cl | 2 | | |
| $HCO_4$ | 2 | | |

the addition of 12 uc of $^{45}$Ca, we harvested the seedlings, prepared samples for counting, and used whole
plants to make radioautographs.

Ca exchange was measured in 2-inch stem pieces
from seedlings grown in $^{45}$Ca. One-half ml. of salt
solutions, placed in tubing attached to the stem
piece, was allowed to flow acropetally through the
piece. The $^{45}$Ca was determined in the eluate.

We grew seedlings for 2 years in 3-gallon crocks
of a soil-sand-peat mixture with which the following
quantities of salts had been thoroughly mixed at
planting time: $KH_2PO_4$, 10.5 g; $Mg(OH)_2$, 4.5 g; $Ca(OH)_2$,
4.5 g; and either $NaNO_3$, 18 g; $NH_4NO_3$, 9 g; or
$(NH_4)_2SO_4$, 15 g. To inhibit the conversion of $NH_4^+$

to $NO_3^-$, 0.06 g of 2-chloro-6-(trichloromethyl)
pyridine per crock was mixed with the N source before
mixing in the substrate.

We prepared radioautographs by exposing the dried
plants to X-ray film for 1 week.  Samples for radioactivity were counted in a planchet-counting system
equipped with a gas-flow detector.

In the controlled-nutrient cultures, we grew each
tree in well drained pure quartz sand in a sunken
55-gallon drum.

All trees were propagated in the fall of 1965 on
MM 26 rootstock with buds taken from one 'York
Imperial' tree.  Differential nutrient treatments were
started on May 17, 1966.

The compositions of the 36 nutrient solutions are
shown in Table 2.  We applied 20 1. of solution weekly
during the growing season of 1966 and 1967, and 3
times every 2 weeks in 1968.

Median leaves from terminals of current season's
growth were used for analyses.  Samples were washed
quickly in detergent solution, rinsed in deionized
water; dried in a forced-draft oven at 70°C; and
ground in an intermediate Wiley mill to pass a 40-mesh
screen.  Representative fruit samples consisting of
the cortex only were freeze-dried and ground as were
the leaves.

Mineral elements in the leaves and fruit were
determined with an "emission" spectrometer as described previously (Baker et al., 1964).  Total N was
determined by the official Kjeldahl-Gunning method.

*Ca movement in seedlings*

Ca moves slowly in apple seedlings (Fig. 1).

TABLE 2. Composition of nutrient solutions used for growing 'York Imperial' apple trees to fruiting in outdoor sand cultures [1]

| Source of N | N | Ca | B | $NH_4NO_3$ mM | $NH_4Cl$ mM | $NH_4OH$ mM | $NaNO_3$ mM | $KNO_3$ mM | $KCl$ mM | $KHCO_3$ mM | $KH_2PO_4$ mM | $K_2SO_4$ mM | $Ca(NO_3)_2$ mM | $CaCl_2$ mM | $CaSO_4$ mM | $Mg(NO_3)_2$ mM | $MgSO_4$ mM | $H_3BO_3$ uM |
|---|---|---|---|---|---|---|---|---|---|---|---|---|---|---|---|---|---|---|
| All $NO_3^-$ | 1 | 1 | 1 | | | | | 1.00 | | | 1.00 | | 1.00 | 1.25 | | 0.50 | 0.50 | 5 |
| " | 1 | 1 | 2 | | | | | 1.00 | | | 1.00 | | 1.00 | 1.25 | | 0.50 | 0.50 | 50 |
| " | 1 | 2 | 1 | | | | | | | | 1.00 | 0.50 | 1.00 | 8.75 | | | 1.00 | 5 |
| " | 1 | 2 | 2 | | | | | | | | 1.00 | 0.50 | 1.00 | 8.75 | | | 1.00 | 50 |
| " | 1 | 3 | 1 | | | | | | | | 1.00 | 0.50 | | 16.25 | 2.28 | 0.50 | 0.50 | 5 |
| " | 1 | 3 | 2 | | | | | | | | 1.00 | 0.50 | | 16.25 | 2.28 | 0.50 | 0.50 | 50 |
| " | 2 | 1 | 1 | | | | 2.00 | 1.00 | | | 1.00 | | 1.50 | 1.25 | | | 1.00 | 5 |
| " | 2 | 1 | 2 | | | | 2.00 | 1.00 | | | 1.00 | | 1.50 | 1.25 | | | 1.00 | 50 |
| " | 2 | 2 | 1 | | | | | 1.00 | | | 1.00 | | 1.50 | 3.75 | | 0.50 | 0.50 | 5 |
| " | 2 | 2 | 2 | | | | | 1.00 | | | 1.00 | | 1.50 | 3.75 | | 0.50 | 0.50 | 50 |
| " | 2 | 3 | 1 | | | | | 1.00 | | | 1.00 | | 1.50 | 11.05 | 2.28 | 0.50 | 0.50 | 5 |
| " | 2 | 3 | 2 | | | | | 1.00 | | | 1.00 | | 1.50 | 11.05 | 2.28 | 0.50 | 0.50 | 50 |
| " | 3 | 1 | 1 | | | | 6.00 | 1.00 | | | 1.00 | | 1.88 | 1.25 | | 1.00 | | 5 |
| " | 3 | 1 | 2 | | | | 6.00 | 1.00 | | | 1.00 | | 1.88 | 1.25 | | 1.00 | | 50 |
| " | 3 | 2 | 1 | | | | 2.25 | 1.00 | | | 1.00 | | 3.88 | | | 1.00 | | 5 |
| " | 3 | 2 | 2 | | | | 2.25 | 1.00 | | | 1.00 | | 3.88 | | | 1.00 | | 50 |
| " | 3 | 3 | 1 | | | | | 0.25 | 0.75 | | 1.00 | | 3.88 | | 2.28 | 1.00 | | 5 |
| " | 3 | 3 | 2 | | | | | 0.25 | 0.75 | | 1.00 | | 3.88 | | 2.28 | 1.00 | | 50 |
| .25$NO_3^-$ plus .75$NH_4^+$ | 1 | 1 | 1 | 0.50 | 1.00 | | | 0.25 | 0.75 | 1.00 | 1.00 | | 0.25 | 1.25 | | 1.00 | | 5 |
| " | 1 | 1 | 2 | 0.50 | 1.00 | | | 0.25 | 0.75 | 1.00 | 1.00 | | 0.25 | 1.25 | | 1.00 | | 50 |
| " | 1 | 2 | 1 | | 1.50 | | | | | 1.00 | 1.00 | | | 6.25 | 1.82 | 1.00 | | 5 |
| " | 1 | 2 | 2 | | 1.50 | | | | | 1.00 | 1.00 | | | 6.25 | 1.82 | 1.00 | | 50 |
| " | 1 | 3 | 1 | | 1.50 | | | | | 1.00 | 1.00 | | | 15.00 | 3.88 | 1.00 | | 5 |
| " | 1 | 3 | 2 | | 1.50 | | | | | 1.00 | 1.00 | | | 15.00 | 3.88 | 1.00 | | 50 |
| " | 2 | 1 | 1 | 1.00 | 2.00 | | | | | 1.00 | 1.00 | | 0.50 | 1.25 | | 1.00 | | 5 |
| " | 2 | 1 | 2 | 1.00 | 2.00 | | | | | 1.00 | 1.00 | | 0.50 | 1.25 | | 1.00 | | 50 |
| " | 2 | 2 | 1 | | 3.00 | | | | | 1.00 | 1.00 | | 0.50 | 2.50 | 2.05 | 1.00 | | 5 |
| " | 2 | 2 | 2 | | 3.00 | | | | | 1.00 | 1.00 | | 0.50 | 2.50 | 2.05 | 1.00 | | 50 |
| " | 2 | 3 | 1 | | 3.00 | | | | | 1.00 | 1.00 | | | 10.00 | 4.34 | 1.00 | | 5 |
| " | 2 | 3 | 2 | | 3.00 | | | | | 1.00 | 1.00 | | | 10.00 | 4.34 | 1.00 | | 50 |
| " | 3 | 1 | 1 | | 2.00 | 2.00 | | | | 1.00 | 1.00 | | 1.00 | 1.25 | | 1.00 | | 5 |
| " | 3 | 1 | 2 | | 2.00 | 2.00 | | | | 1.00 | 1.00 | | 1.00 | 1.25 | | 1.00 | | 50 |
| " | 3 | 2 | 1 | | 2.00 | 3.50 | | | | 1.00 | 1.00 | | 1.00 | | 1.60 | 1.00 | | 5 |
| " | 3 | 2 | 2 | | 2.00 | 3.50 | | | | 1.00 | 1.00 | | 1.00 | | 1.60 | 1.00 | | 50 |
| " | 3 | 3 | 1 | | 2.00 | 3.50 | | | | 1.00 | 1.00 | | | 2.50 | 4.79 | 1.00 | | 5 |
| " | 3 | 3 | 2 | | 2.00 | 3.50 | | | | 1.00 | 1.00 | | | 2.50 | 4.79 | 1.00 | | 50 |

[1] Micronutrients were supplied in the following micromolar concentrations of the indicated salts: FeEDTA,117; $(NH_4)_6Mo_7O_{24}$,0.7; $ZnSO_4$, 7.6; $CuSO_4$, 3.9; and $MnSO_4$, 9.

after 3 days                    after 7 days

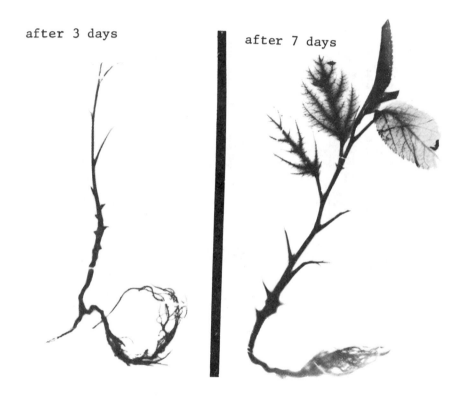

FIG. 1a. Movement of Ca into apple seedlings.
(3 and 7 days after introduction of
$^{45}$Ca to the roots.)

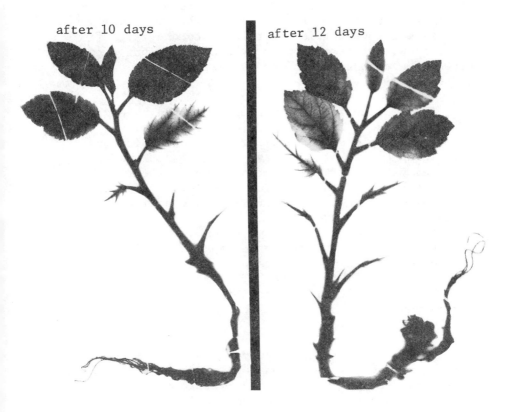

after 10 days            after 12 days

FIG. 1b. Movement of Ca into apple seedlings.
(10 and 12 days after introduction
of $^{45}$Ca to the roots.)

Only after 3 days had $^{45}$Ca reached the tip of a 12-
inch plant, and not until after 7 days had the laminae
of the youngest developing leaves accumulated significant quantities of the $^{45}$Ca. Accumulation in the
older leaves slowly progressed basipetally, first
appearing in the petioles, then the midribs and main
veins, and finally in the leaf blades. Even after 12
days, the petioles and midribs only of the oldest
leaves had taken in any $^{45}$Ca. If Ca is moved by the
transpiration stream, as pointed out by Wiersum (1966),
its rate of movement is not related to the rate of
water flow, but it must move by exchange along a continuous "exchange column" as suggested by Bell and
Biddulph (1963).

That Ca can be moved through the stem by exchange
was demonstrated by passing Ca, Mg, or Ba salts
through stem pieces of seedlings grown in solutions
containing $^{45}$Ca. All of these divalent cations
replaced Ca in the stem in proportion to their
molarity in the exchange solution (Fig. 2). Solutions
of $H_3BO_3$ or of $NH_4^+$ salts, or water alone did not
release Ca.

We determined the exchange capacity of lignin for
Ca and found that Ca combines with the lignin molecule
in easily exchangeable form. The association of $^{45}$Ca
with highly lignified tissues (Fig. 1) indicates that
lignin may be an important exchange site for the movement of Ca in apple.

Ca moves by exchange and does not move in proportion to the transpiration rate. Its initial movement into developing rather than mature apple tissues
(Martin, 1967) must depend on metabolic removal from
exchange sites (Bell and Biddulph, 1963) by tissues in
proportion to metabolic requirements.

Increasing their metabolic activity with sprays
of kinetin can move $^{45}$Ca into old leaves (Fig. 3).

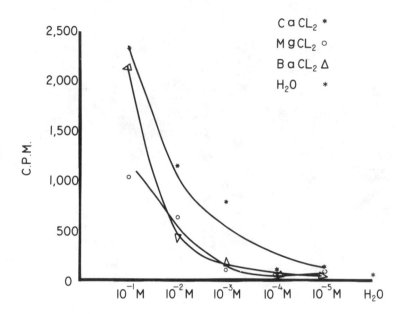

FIG. 2.   Exchange of $^{45}$Ca from stem pieces of
          apple seedlings.  (cpm indicates $^{45}$Ca
          eluted from stems with 0.5 ml of salt
          solution of indicated molarity.)

Concentrations of $2.5 \times 10^{-4}$ M. kinetin and benzyl-
adenine increased $^{45}$Ca movement into the base of old
leaves by 59.2 and 43.5% and into their tips by 78.2
and 19.9%, respectively, as compared to unsprayed
check leaves.

We also increased the movement of $^{45}$Ca into the
petioles and midrib of apple leaves with foliar sprays

FIG. 3.  Ca transport into mature apple leaves
        sprayed with kinetin.  K, sprayed with
        500 ppm kinetin; U, unsprayed control.

of $H_3BO_3$ (Table 3).

Source of N supplied to the soil also affected Ca
transport (Table 4).  When N was supplied as all $NO_3^-$,
the Ca distribution followed the "normal" pattern,
high in basal and low in terminal leaves; when all N
was supplied as $NH_4^+$, Ca concentration decreased in
basal and increased in terminal leaves.  Distribution
in plants supplied with $NH_4NO_3$ fell between these
extremes.  Similar results were obtained on young
seedlings in solution cultures receiving different

TABLE 3

Effect of Four Sprays of $1.6 \times 10^{-3} M H_3 BO_3$ on Ca Transport in Apple Leaves. Values Given as % of Length of Petioles and Mid-Rib Occupied by $^{45}Ca$ on Radioautographs

| Ca level | Mid-shoot leaf | | Basal leaf | |
| --- | --- | --- | --- | --- |
| | -B | +B | -B | +B |
| Low | 62.8 | 92.3 | 61.1 | 96.6 |
| Medium | 62.3 | 85.8 | 61.7 | 88.4 |
| High | 49.7 | 77.1 | 58.6 | 80.0 |

TABLE 4

Ca Content of Apple Leaves Receiving Different Sources of N*

| | $NO_3$ | $NH_4NO_3$ | $NH_4$ |
| --- | --- | --- | --- |
| | Ca % dry weight | | |
| Terminal leaves | .143 d | .323 c | .403 c |
| Basal leaves | .923 a | .877 a | .620 b |

* In soil; soil was mixed with 2 chloro-6-(trichloro-methyl) pyridine. Mean values followed by the same letter are not significantly different at the .05 level of significance.

ratios of $NO_3^-$ to $NH_4^+$ N (Fig. 4). Spraying with a 1%
solution of $(NH_4)_2SO_4$ also reduced the Ca accumulation
in the older leaves (Table 5).

## Ca accumulation in leaves and fruit of bearing trees

Ca content of leaves and fruit (flesh plus peel)
from Ca-deficient "York Imperial" trees and from trees
classed as Ca-sufficient on the basis of leaf Ca are
shown in Fig. 5. The general characteristics of the
curves are similar to those of Rogers et al. (1953,
1954). Interpreted on the basis of the information on
movement and accumulation of Ca already presented, they
permit a number of deductions.

Median leaves on terminal shoots are older at each
successive sampling date during the season. High metabolic activity during the period of leaf expansion promotes movement of Ca into the laminae of the leaves.
After maturity, increasing lignification of the vascular
tissues provides an expanding reservoir for accumulation
of Ca. Once terminal growth ceases, metabolic demand
of developing tissues is reduced and Ca build-up depends
mainly on accumulation on exchange sites. This, in
turn, can be influenced by moisture relations. Low
soil moisture restricts Ca uptake and availability in
the transpiration stream. Adequate soil moisture and
high transpiration promote movement of Ca to exchange
sites. Low soil moisture and high transpiration might
induce movement of water from the fruit and consequently
remove some of the Ca from the exchange sites in fruit
(Wilkinson, 1968).

TABLE 5

Effect of $(NH_4)_2SO_4$ and Urea Sprays* on
Movement of $^{45}Ca$ into Mature Apple Leaves

|  | 6th leaf from apex | 7th leaf | Stem between 6th and 7th leaves |
|---|---|---|---|
|  | cpm/cm$^2$ of area | | cpm/0.1 g tissue |
| Control | 212 c | 267 c | 1211 a |
| $NH_4^+$ sprayed | 89 d | 61 d | 839 b |
| Urea sprayed | 213 c | 217 c | 1022 a |

* Sprayed with 0.5% $(NH_4)_2SO_4$ or 0.5% urea on 1st and
4th day after introduction of $^{45}Ca$ to the roots.
Leaves were counted for radioactivity 10 days after
the first spray. Mean values followed by the same
letter are not significantly different according to
Duncan's Multiple Range Test.

The leveling off of the percent of Ca in the fruit
during mid-summer, accompanied by a rapid increase in
fruit weight, represents an actual increase in Ca per
fruit. This agrees with the data of Rogers et al.
(1954) and Wilkinson (1968).

Reduction in leaf Ca between July 21st and August
4th cannot be explained from the limited data available. Rate of shoot and fruit growth, moisture conditions, or combinations of these and perhaps other
factors may be involved. Further work is planned to
clarify this point.

In view of the effects on Ca movements of $NH_4^+$ and
B, their effects on leaf and fruit Ca in bearing trees
in sand cultures are of interest.

FIG. 4.  Transport of $^{45}Ca$ into mature leaves of apple
         seedlings receiving various sources of N
         nutrition.

         A) all $NO_3^-$; B) 1/3 $NH_4^+$ plus 2/3 $NO_3^-$; C) 3/4
         $NH_4^+$ plus 1/4 $NO_3^-$.

        Interactions between source of N and level of Ca
on leaf K, Ca, and Mg (average of three years analyses)
are shown in Fig. 6.  At all levels of Ca, leaves of
trees receiving $.75NH_4^+$ plus $.25NO_3^-$ N were higher in
Ca than were leaves from trees receiving all $NO_3^-$ - N
(Fig. 6).  Only at the low level of Ca was leaf Mg
higher under $NH_4^+$ than under $NO_3^-$ nutrition.

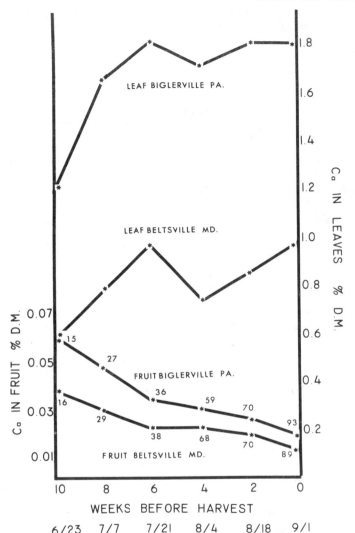

FIG. 5.   Seasonal changes in the calcium content of
          leaves and fruits of "York Imperial" apple
          trees from a calcium-sufficient orchard
          (Biglerville, Pa.) and from a calcium-
          deficient orchard (Beltsville, Md.).  Fig-
          ures accompanying fruit data are average
          weights per fruit in grams.

90

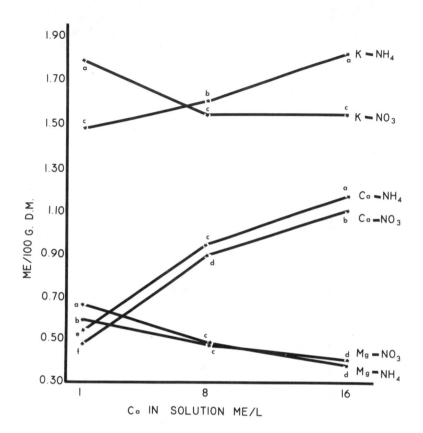

FIG. 6. Interactions of levels of calcium with sources
of nitrogen on the content of potassium, calcium, and magnesium of leaves of "York Imperial"
apple trees growing in controlled-nutrient
cultures. All data are averages of 3 levels of
nitrogen supply.

Level of Ca affected leaf K differently, depending
on source of N (Fig. 6). With all $NO_3^- - N$, increasing
Ca decreased leaf K but had the opposite effect when N
was supplied as $.75NH_4^+$ plus $.25NO_3^-$. Since the K/Ca

ratio has been considered a significant factor in the
incidence of apple disorders associated with low Ca,
this effect of source of N on K could have important implications.

The effect of B on leaf Ca as measured by median-
shoot leaf samples (Table 6) appears to be more complex than is indicated by the short term experiments
with B sprays (Table 3).  The effect of B level on leaf
Ca was influenced by interactions between source of N
and level of Ca (Table 6).  Except for the slight increase in Ca associated with increased B at the intermediate level of Ca with all $NO_3^-$ - N in 1966, the B
effects were consistent and became more pronounced
each year.  At the intermediate level of Ca with $NO_3^-$,
leaf Ca was higher with the higher B supply.  At the
high level of Ca with $NH_4^+$ - N, however, the effect of
B was reversed.  Until we can determine the mechanisms
responsible for the $NH_4^+$ - N and the B effects on Ca
accumulation, these interactions must remain unexplained.

Ca did not accumulate proportionately in leaf and
fruit tissues (Table 7).  Leaf Ca was 65 percent higher
at the second level of Ca as compared with the lowest
level, and 24 percent higher at the third as compared
with the second level.  Comparable increases in fruit
Ca were 23 and 14 percent, respectively.  Leaf symptoms
of Ca deficiency appeared in 1968 on trees in all low-
Ca treatments.  Characteristic symptoms associated with
low Ca also appeared early on the fruit, and progressed
to severe fruit breakdown.  No leaf symptoms of Ca deficiency developed on trees that received intermediate
or high levels of Ca; but the fruit symptoms identical
to those that had appeared early on the low-Ca treatments developed later in the season on the fruit of the
trees that received the intermediate level of Ca.  Thus,
a continuous supply of Ca sufficient to prevent leaf
symptoms of Ca deficiency was not sufficient to prevent
Ca deficiency in the fruit.

TABLE 6

Effects on Source of N and Levels of Ca and B on Percent Ca
in Apple Leaves —

| Source of N | Level of Ca | Level of B | 1966 % | 1967 % | 1968 % | All Years % |
|---|---|---|---|---|---|---|
| $NO_3^-$ | 1 | 1 | .42 qr* | .43 qr* | .58 nop* | .48 h* |
| $NO_3^-$ | 1 | 2 | .44 qr | .38 r | .64 mno | .48 gh |
| $.25NO_3^-$ plus $.75NH_4^+$ | 1 | 1 | .54 op | .42 qr | .67 lmn | .54 fg |
| $.25NO_3^-$ plus $.75NH_4^+$ | 1 | 2 | .50 pq | .43 qr | .71 klm | .55 f |
| $NO_3^-$ | 2 | 1 | .78 ijk | .73 klm | 1.14 de | .89 e |
| $NO_3^-$ | 2 | 2 | .89 h | .71 klm | 1.06 ef | .89 e |
| $.25NO_3^-$ plus $.75NH_4^+$ | 2 | 1 | 1.02 fg | .69 klm | .95 gh | .89 e |
| $.25NO_3^-$ plus $.75NH_4^+$ | 2 | 2 | 1.07 def | .76 jkl | 1.12 def | .98 d |
| $NO_3^-$ | 3 | 1 | 1.10 def | .85 hij | 1.26 bc | 1.07 c |
| $NO_3^-$ | 3 | 2 | 1.17 cd | .85 hij | 1.38 a | 1.13 b |
| $.25NO_3^-$ plus $.75NH_4^+$ | 3 | 1 | 1.34 ab | .92 h | 1.34 ab | 1.20 a |
| $.25NO_3^-$ plus $.75NH_4^+$ | 3 | 2 | 1.31 ab | .88 hi | 1.17 cd | 1.12 bc |

* Values not followed by a common letter are significantly different at the
0.05 level of confidence.

TABLE 7

Effects of Level of Ca Supply on Leaf
and Fruit Ca, 1968.  Average of All
Sources and Levels of N and Levels of B

|       | Leaves<br>% D.M. | Fruit<br>% D.M. |
|-------|------------------|-----------------|
| Ca1   | 0.65 c*          | 0.0068          |
| Ca2   | 1.07 b           | 0.0084          |
| Ca3   | 1.29 b           | 0.0096          |

* Values not followed by a common letter are
  significantly different at the 0.05 level of
  confidence.

CONCLUSIONS

The interpretation of tissue content of Ca in terms
of Ca status is fraught with more pitfalls than that of
any other major nutrient.  Enhancement of the Ca content
of specific plant tissues, particularly certain areas of
the fruit, also presents challenging problems.  Perhaps
little that we have presented here is new.  Our desire
has been to focus attention on the unique aspects of the
mechanisms of Ca translocation and accumulation, and to
point out how some environmental and nutritional variables may influence these mechanisms.  We hope that the
integration of all these considerations may lead not
only to increased usefulness of tissue analyses for Ca,
but to means of promoting the movement of Ca into those

organs and tissues that are particularly vulnerable to
disorders associated with Ca stress.

REFERENCES


1.   BAKER, D.E., GORSLINE, G.W., SMITH, C.B., THOMAS,
     W.I., GRUBE, W.E. and RAGLAND, J.L. (1964)
     Technique for rapid analyses of corn leaves for
     eleven elements. *Agron. J. 56*: 133-136.

2.   BELL, C.W. and BIDDULPH, O. (1963)  Translocation
     of calcium.  Exchange versus mass flow. *Plant
     Physiol. 38*: 610-614.

3.   FAUST, M. and SHEAR, C.B. (1968)  Corking dis-
     orders of apples.  A physiological and bio-
     chemical review. *Bot. Rev. 34*: 441-469.

4.   MARTIN, D. (1967)  $^{45}$Ca movement in apple trees.
     Experiments in Tasmania 1960-1965. *Field Stat.
     Record, Div. of Plant Industry, C.S.I.R.O., Ho-
     bart, Tasmania 6*: 49-54.

5.   PERRING, M.A. (1968)  Mineral composition of apples.
     VIII: Further investigations into the relation-
     ship between composition and disorders of the
     fruit. *J. Sci. Fd. Agric. 19*: 640-645.

6.   ROGERS, B.L. and BATJER, L.P. (1954)  Seasonal
     trends in six nutrient elements in the flesh of
     Winesap and Delicious apple fruits. *Proc. Amer.
     Soc. Hort. Sci. 63*: 67-73.

7.   ROGERS, B.L., BATJER, L.P. and THOMPSON, A.H. (1953)
     Seasonal trend of several nutrient elements in
     Delicious apple leaves expressed on a percent
     and unit area basis. *Proc. Amer. Soc. Hort. Sci.
     61*: 1-5.

8.    WIERSUM, L.K. (1966)  Calcium content of fruits
      and storage tissues in relation to the mode of
      water supply.  *Acta Botanica Neerlandica 15*:
      406-418.

9.    WILKINSON, B.G. (1968)  Mineral composition of
      apples.  IX: Uptake of calcium by the fruit.
      *J. Sci. Fd. Agric. 19*: 646-647.

*Questions to Dr. Shear*

CHAMEL:  Pensez-vous que le déplacement du calcium
des feuilles vers le fruits soit un phénomène important?
On sait que le calcium se déplace très bien des racines
vers le sommet de la plante.  Les études avec des éléments
radio-actifs ont montré par contre, que le déplacement du
calcium à partir des feuilles est très faible.  Ce comportement du calcium, justement dans son déplacement à
partir des feuilles bien ressemble à celui des autres
alcalino-terreux, comme le strontium, par exemple.  Alors
qu'avez-vous fait vous-même des expériences sur ce déplacement du calcium à partir des feuilles?

SHEAR:  We have tried to move calcium from leaves but this
is a very difficult thing to do.  In the first place, in
order to move it from the leaves into the fruit you have
to have the fruit demanding it.  There has been some work
done by others which indicated that in periods of moisture
stress calcium moves from the fruit into the leaves.  But
I do not think that there has been any work done to show
that you can move calcium out of the old leaves into the
new ones.

BAR-AKIVA:   In discussing calcium nutrition, we should
consider the role of boron.   Generally we found, at least
in citrus, that with calcium intake we have an accumulation of boron.   On the toxic side, or near the toxic
side of boron we have a decrease of calcium.   One of the
treatments we can use to improve the health of the tree
in case of boron toxicity in citrus is to find means of
increasing calcium.   Sometimes you may have the same
level of boron in two trees.   One is suffering from
toxicity because the calcium:boron ratio is low, and in
the second one high and therefore healthy.

SHEAR:   You will notice that the boron sprays increased
the calcium in the whole leaf blade; (which indicates
that when we went from a boron deficiency to a boron
sufficiency we were apparently affecting the metabolic
rate of the leaves).   This may possibly have been
brought about through some effect on hormones, since
some of the work has shown that boron and N.A.A. sprays
combined can increase the movement of calcium into apple
fruit.   So it would appear that this is a metabolic
effect due to increasing boron to a sufficiency level.

WOLF:   Does the addition of boron actually increase the
calcium in the fruit, or was this just the calcium in
the leaves?

SHEAR:   In this case it is the calcium of the leaves.
But I showed that calcium in the fruit has been increased by means of boron.   Also, in the last month
papers have come out by Frank Hewetson and others working
under Dr. Childers, showing that they have been able to
increase it in the fruit.

CLEMENTS:   I am interested in the comment that you made
made that since calcium going up in the xylem has to be
on a base exchange medium, whatever goes up as such has
to go up through the phloem.   I think Biddulph has shown
that fixation in the various xylem sites is not a very
permanent one at all, and that calcium continues to
move upward in the xylem.

SHEAR:   I did not intend to give the impression that
the calcium did not move in the xylem.   I feel that
the calcium that moves through the phloem has to do so
by activated movement, whereas this exchange movement
of calcium takes place in the xylem.

Most of the calcium that is in the leaf is in the
conducting elements in the xylem.   There is very little
calcium outside them.   Of course, you cannot tell from
radioautograms like these, because that is not a quantitative measure;  but the calcium that is in the
petiole probably does not move out very readily, and
within the main body of the leaf, moves out with difficulty.

ROUTCHENKO:   J'ai été très intéressé par vos observations concernant le rôle de NAA, en tant que semble-
t-il véhicule du calcium entre des feuilles de niveau
différent.   Est-ce que vous reliez ce rôle, un rôle de
transport ou pêut-etre à une action sur la composition
analytique organique qui sont pour une bonne part re-
sponsables due bloquage, notamment l'acide oxalique?
Quel est, d'après-vous le rôle de $NH_4^+$ dans les différences que vous constatez dans la transmission entre
les feuilles?

SHEAR:   We do not know what the role played by ammonium
is; we have some working theories which may be way off.
We know that you get a great difference in the composition of all plants if you supply ammonium or nitrate

nitrogen, and there is some work to show that there is a difference in the amino acid composition of roots with ammonium versus nitrate. It is very possible that these amino acids may be chelating agents that help to move the calcium when you have ammonium nitrogen. That is the only thought that we have at this time.

# GENERAL DISCUSSION

## *The mathematical approach to fertilizer requirements*

KAFKAFI:  I would stand for a mathematical approach,
but not for using it literally.  By using mathematics
we are trying to understand the system and to apply
it more accurately.  But assuming only one variable
when all the others are kept constant, we will never
get to the point where we can reach the maximum use.
I would like to see the approach to the maximum
potential use put down in terms of genetics, light
conditions, mineral contents and water supply.  And
then we should, when knowing all these factors, calculate, on the basis of light intensity and energy
produced in a certain area what we can expect for a
maximum production. Then we should try to use a
fertilizer to reach this amount, no matter whether it
will be any accepted law or not.

DE WIT:  You all know that when one puts dry matter
used and yield versus nutrient, one gets, as long as
the nutrient is deficient, a straight line.  The
moment we have another factor which controls the yield
we will get deviations from that line, and we know
that in order to correct it we should look for another
element which is capable of doing this.  This means

99

that we have a linear function.  If we know this (per
unit) for each element, we might reach a point where the
maximum yield or dry matter yield is controlled by either
light, $CO_2$ content, or temperature; things which we can--
not control unless we are working in a controlled environment.  So I would like first to calculate the maximum and
then to see whether we are in a place where a change of
nutrient status can help us any more, or if we have
reached the optimum in our environment.

These were analyses done on fertilizer experiments.
Essentially, as far as we understood, there are three
fertilizer levels.  Now, in the midst of these observations that were used there were three unknowns.  So
there isn't much to be gained really by using the
Mitscherlich equation.  No degrees of freedom are left
to test whether the equation will hold or not.  In
general I would say that the point made by Mitscherlich
was of the constancy of the working factor, and that it
is constant and independent of the growth conditions.  Of
course, it hasn't been proved, but it was nicely marketed
at that time.  I don't see much point in the whole thing.
Moreover, I miss in this mathematical approach the relationship between the cost of fertilizer and the cost of
the produce, or the cost of application of the fertilizer.
This may give you a completely wrong idea of the efficiency
of the use of fertilizer.  I think we are, in many agricultural operations in the situation that we have to
decide whether we should apply enough fertilizer or
whether we should apply no fertilizer at all; for in
this case we save the cost of application, and that may
be the main cost.  Fertilizer is just pocket money.

MALAVOLTA:  As you know, the approach presented by
Dr. Kafkafi has been inserted.  In the particular case of
corn, there are some theoretical considerations showing
that one could get, if there were no limiting factors,

yields of over 15 tons per hectare. And as far as I
know, the highest recorded yields of this corn are
about 15,000 lbs per hectare. Our philosophy is to
restrict ourselves to a small number of variables,
otherwise we would run into trouble. The experiments
would be too complicated and they are also very difficult to interpret. Concerning the remarks by
Dr. de Wit, I would like to point out that we have
been using in Brazil for 20 years the Mitscherlich
equation - not the limiting factor concept. We do
know that C is not constant, so one has to calculate
the C value every time experiments are run. Concerning this multitude of curves and the possibility of
simply drawing curves by hand, I would like to point
out that in certain cases equations would keep closer
to the experimental data. I realize that we are using
only three levels of fertilizers in most of the experiments, but not in all of them. We get a considerable amount of data by analysing results of the
experiments according to soil types and climatic conditions. This gives a very good basis for the statistical analysis. The general experience in Brazil
has been rather satisfactory by using the old
Mitscherlich equation, not the whole concept.

ROUTCHENKO: Tout le monde sera certainement d'accord
pour dire l'expression mathématique de nos expériences
est une nécessité absolue. Mais il est nécessaire que
des formules mathématiques qui pourraient être
établies soient basées sur des données précises, et
complètes, et c'est là certainement que la difficulté
commence. D'autre part, nous nous basons généralement
sur des données purement expérimentales, et, n'attachons
peut-être pas assez d'importance aux données fondamentales
de la nutrition. Et, j'aimerais poser la question à
cette assemblée, pour savoir si vous ne croyez pas qu'il
serait nécessaire de donner une définition biologique

au besoin de la plante dans un élément donné.  C'est-
à-dire, établir des critères objectifs, et indépendants
des conditions locales particulières dans lequel le
végétal se développe, pour juger de la manière dont
l'alimentation dont ils bénéficient correspond à ses
besoins.  En ce qui nous concerne, nous définissons ces
besoins comme la possibilité que le végétal a dans les
conditions données d'utiliser les éléments, soit par voie
métabolique, soit par voie de transfert dans les organes
en formation.  Je crois qu'il a là une possibilité de se
libérer des indices établis expérimentalement, et qui de
ce fait ne sont valables que pour des conditions données.

NIELSEN:  If we have a number of points available we
may use the Mitscherlich equation and we often use it.
We are sometimes able to use the Mitscherlich equation
by putting in another constant, but it will be dead and
most of the time it will not show the curve line;  and
we might run into two or more constants and so on.  But
I think that there exists a theory against the
Mitscherlich equation according to which it only fits
in one part of the curve and not really well in the
other part of the curve.  I think that there is only one
advantage to the Mitscherlich equation;  it excludes the
personal element.  This is not so important.  We have to
use our experience when we draw a line, and I still
think that there is a wonderful faculty in man to draw
it the right way.  Such a curve is much easier and
better understood than an equation.

BOWEN:  In all these discussions about equations, there
seems to be an aura of exactness about the whole thing
in terms of what you get and what your response is.  I
would like to ask just how much precision do you expect
for any kind of fertilizer application in view of the
variability imposed by environmental conditions, such
as those which Dr. Clements raised.  For example, I
think you will find that while a certain type of curve

may reflect the response in a particular soil, depending on the particular seasonal conditions, your response
to some application will fit only a limited part of that
curve.

CLEMENTS:  While I don't have much regard for the
Mitscherlich equation, I certainly think we should try
to express our systems mathematically.  The point just
made by the previous speaker shows us a way; that is,
what is the maximum yield in a particular place?
I don't think that growing plants in pots in a greenhouse is likely to give you accurate results as far as
yield results are concerned, largely because of the fact
that you have interfered with the light and you have
certainly restricted the root development.

In answer to what kind of accuracy you can expect,
we get correlations of about 0.92.  The way we do it
is this; on all of our sugar fields we take periodic
plant samples for analysis.  Also we record the sunlight,
the input of radiation, maximum and minimum temperature,
as well as the complete analysis for all the elements.
When we are dealing with low pH's, we also include root
analysis because they generally are better indicators
for toxic elements.  When we subject all these data -
anywhere from five to ten thousand sets - to a multiple
regression analysis, we can get this high correlation.
We can express the results which becomes applicable on
the basis of the best yield obtained in an outline plan-
tation to a current situation.  It is of interest that
the dominant factors do not include any of the nutrient
elements.  They are: radiation, maximum and minimum
temperature, the age of the plant at which you are
making your growth estimate and tissue moisture.  If
there is a deficiency of any element, you do not get
normal tissue moisture up where it belongs.  So then you
end up with a fairly good integration.  The trouble

with the multiple regression equation is that you are
dealing with factors that are added to one another. We
are at the present time working to change this, so that
if we actually get a single factor that is dropping below the critical level so that it is actually a zero
factor, we should still be able to use the results. The
nutrients are not by themselves very important unless
they are deficient, and this is reflected in the moisture
level.

BOWEN: Dr. Clements' points make my point too. There
can be obtained a very high degree of correlation on
hindsight by taking the weather conditions, radiation,
and so on into account in that particular season. I'm
in favor of trying to express these things mathematically.

Nutritional conditions can only account for a certain percentage of your variability in a quite expensive
series of experiments. People think they're doing very
well if they can explain 60 to 70% of the variability in
yield on soil nutritional characteristics, and it's for
this reason that we need these curves. But we shouldn't
expect a 5% accuracy in prediction.


*Chemical form of nutrients in plants.*

BOULD (SESSION LEADER): I would like to come back to
Dr. Routchenko's question concerning the chemical form
of nutrients in plants, particularly in relation to tissue
analysis. With leaf analysis people originally used
total analysis; now there seems to be a change and
workers are more interested in function and form and so
on.

BIELSKI: We have been working (free of soil) with little
water plants growing in solution, and we have noticed
quite a definite pattern with amino acids and with phos-

phate, and also with carbohydrate, in that we impose a
deficiency in one, we very often get the appearance of
a super-sufficiency in another.

For example, if we have imposed sulphur deficiency
or phosphorus deficiency, or if we grow our plants in
darkness they will grow on carbohydrate.  Then we get
the appearance in the amino acids of very high levels
of arginine, glutamine, asparagine, and conversely.
This is characteristic also of supplying nitrogen at
very high levels.  Under conditions of say radiation
damage, or growth in darkness, or nitrogen deficiency
we find the ratio of inorganic phosphate to total phosphate rising, so that in the case of phosphorus the
thing that characterizes an oversupply of phosphorus or
an undersupply of something else is a high level of
inorganic phosphate in the tissue.  I suspect that there
are other things that will follow this pattern.

ROUTCHENKO:  Plusieurs orateurs nous ont fait part de
leurs expériences extrêmement intéressante.  Je pense
qu'il y aurait peut-être intérêt d'aborder un sujet
plus fondamental, c'est à dire celui des bases même
du fondement du diagnostic.  Nous ne pourrons pas nous
limiter a constater simplement un état nutritif, nous
devons aller plus loin et comprendre le mécanisme qui
conduise à la production plus ou moins importante d'une
plante et d'une recolte.

J'ai insisté sur la définition des besoins du
végétal.  Ces definitions permettent dans une certaine
mesure de se passer du nouveau critique établi
expérimentalement, c.a.d. la plante analysée fournit
en grande partie elle-même ses propres critères.
Lorsque nous constatons un certain déséquilibre
d'alimentation, nous trouvons l'accumulation de
certains éléments et au contraire des niveaux

excessivement bas d'autres.  Ce sont ces méchanismes qui
permettent de juger effectivement comment cette plante
là, la plante que l'on analyse, est alimentée contre ses
besoins métaboliques.  C'est-à-dire, intégrant toutes
les conditions écologiques dont plusieurs auteurs ont
parlé, et qui en grande partie déterminent des
possibilités d'utiliser tel ou tel niveau alimentaire
dans le sol.  Il me semble, que dans un assemblée comme
la vôtre, nous pouvons peut-être justement dépasser,
comme les organisateurs de colloques ont souhaité, le
type d'une expérience determinée, et d'une résultat
extrêmement intéressant.

ARATEN:  The chemical form of various fertilizers is of
great importance, so that we should suggest that all
publications on fertilizer experiments should not just
mention NPK, but should mention the form.  Even when
we are concerned with the cation, its accompanying
anion should be given because that may be of influence.

*Factors affecting movement of calcium.*

DE WAARD:  Le sujet de la migration du calcium dans la
tomate, est une question qui a intéressé beaucoup les
producteurs de tomates sous verre, puisque il est lié à
la maladie du "Blossomend Rot."  Je parle des travaux
néerlandais, qui ont montré qu'il y avait graduation
très grande avec l'intensité de la transpiration.
Lorsque la tomate transpirait mal, il y avait cette
maladie, lorsqu'elle transpirait beaucoup, elle ne
l'avait pas.  Ils ont établi que le calcium venait
plûtot des vaisseau du bois, de la sève montante, que
de vaisseaux du liber, de ce que l'on appelait autrefois,
la sève elaborée.  Ceci est très important, et se relie
à l'information qui a été donnée, en disant justement,
que le calcium migrait très difficilement de la feuille,
et pour aller dans le fruit, il faut qui'il vienne de la
racine.  Contrairement la composition du milieu est
important et les processus physiologiques comme la
transpiration sont extrêmement importants.  Je ne sais
pas si la composition des anions est important également.
Par exemple, on pourrait se poser la question de
savoir si la présence de chlore, par exemple, anions
non métabolises aiderait à la migration du calcium

dans le fruit.

CAMPBELL:  We have had problems for many, many years
getting enough calcium into tomato fruits.  We know that
if you have less than 0.2% Ca in the tomato fruit you
can expect blossom end rot, which is really calcium deficiency, and also severe cracking.  It was found some
years ago that if you put a foliar application of calcium
on the leaves, you did not get much calcium into the
fruit.  But it was the calcium that landed on the fruit
that did the improvement.

We have a problem in Manitoba in relation to
blossom end rot tomatoes.  I think the reason is that, as
in Australia, we have soils which have very high magnesium levels.  In other words, if you look at the ratio
of the calcium to the magnesium in some cases, especially
at the lower levels, we're getting a one-to-one ratio
instead of what would be close to the desirable, a six-
to-one ratio.  We also have very high levels of potassium.
These two factors, plus the ammonium ion largely inhibit
the uptake of calcium.

SHEAR:  The problem is universal with many fruits.
We have the problem with pears.  We have three problems
of this kind with apples; one is an early deficiency
of calcium that we call corking, which is a deep seated
thing that is initiated perhaps six weeks to two months
after full bloom, and then we have the late-occurring
bitter pit.  The same problems exists in all these fruits;
it's some way to get the calcium into the fruit; it just
doesn't go in.  It is a much simpler matter when you are
dealing with a plant like celery or lettuce, where you're
getting it into the leaves.  The only suggestion that I
have is that we try to keep the calcium in solution on
the fruit for a long enough period of time that it has a
chance to get in there.

I don't know when the period of maximum calcium
uptake is in tomatoes, but we know from much of the
work that has been done by Wilkinson and Perring
for example, that the calcium goes into the fruit very
rapidly during the early three or four weeks after full
bloom. From then on, as the fruit expands, we get a
big decrease in calcium percentage and very little increase in total fruit calcium. So the problem is to
know when the metabolism of the fruit would be of the
greatest help in absorbing calcium, and then to put the
calcium on the fruit long enough to have a chance to
work on it and take it in.

WEHRMANN: This kind of calcium deficiency we find not
only in fruits like tomatoes or apples, but we have it
also in the inner leaves of cabbages in the leaves
which have a low transpiration rate.

GREENHAM: I believe it has been suggested that the
production of oxalic acid is increased by the nitrates
compared with ammonium, and the accumulation of
calcium oxalate might be associated with this difference observed in calcium movement when comparing nitrate and ammonium supply.

SHEAR: We used ammonium versus nitrate nitrogen
nutrition and found a very high level of oxalic acid
in the tung tree with ammonium, in all portions of its
stem and leaves, and an extremely low level with the
nitrate treatment. Tung is a peculiar plant which made
excellent growth from seed in deionized water supplied
with analytical grade salts without any calcium, as
long as I used only ammonium nitrogen.

I would assume that there was no calcium oxalate
present because no calcium was supplied, so apparently
with ammonium there was no oxalic acid being produced

in the plant.  Now what was taking the place of the
other essential functions of calcium I don't know.

SHEAR:  It is possible that magnesium could de-
activate oxalic acid as well as calcium, but what was
taking the place of calcium in pectin, for example?
Under field conditions with tung we have been able to
get yield responses with ammonium nitrogen where we
had calcium deficiency; where we got reduced yield
when we put on nitrate nitrogen, and where it was
necessary to put on calcium with the nitrate in order
to get a response.

INGESTAD:  When we are talking about calcium, we
should not forget about the cation balance as a whole.
It is generally referred to as an ion-antagonism in
uptake, but certainly these antagonistic processes
are also going on within the plant at different levels.
Tomato plants are to some extent potassium accumu-
lators, and it is not easy to say if the general problem recognized in the field is a calcium deficiency
problem rather than a cation balance problem.

SHEAR:  I had a paper at the Brussels' meeting on the
calcium-magnesium balance in tung, and I showed there
that calcium and magnesium can, to a certain extent,
replace one another in non-essential functions, so
that if you have a deficiency of total base either one
may be capable of alleviating the deficiency; but
only with the non-essential functions of the individual elements.  I don't think anyone will contradict
the important interactions and it is very true that
potassium will depress the calcium, but in most plants
it will depress magnesium more rapidly than it will
depress calcium.

Here we get back into the mathematical approach.
Twenty five years ago I thought someone would be able

to come up with a mathematical presentation. But even
since computers have taken over, I am still not so
sure that we're not going to have to use a little bit
of what the Lord gave us to think with along with
the mathematical interpretations.

LECOEUILHE: In the case of citrus, when you raise the
concentration of potassium or calcium in the leaf you
decrease the concentration of other cations, so that
you may decrease the total sum of the cations. I think
that in this case, you have to consider the different
forms of calcium.

SHEAR: In my USDA Technical Bulletin in 1954 on calcium,
magnesium, potassium ratios in tung leaves, you'll find
a three-dimensional triangular plot of the total cations,
showing that as you reduce the concentration of magnesium
or potassium, you decrease the total cation; but as you
increase the calcium, you get a very high level of total
cations. So the balance of the total cations in the
leaf depends greatly on the ratio of the three bases.

*The application of pot data to field conditions.*

BOULD (SESSION LEADER): I have used cultures myself,
and as long as one knows what one is doing and uses the
treatments simply to bring about differences in growth
and in nutrient composition so that one can relate
composition of specific parts to the yield of the crop,
then this to me seems reasonably legitimate.

*Summarizing this session,* it seems to me that very little
new has come out since we met last. I think that workers
should be a little more concerned with the timing of
sampling. If you are sampling apple leaves, possibly
for nitrogen, you have to take into account the nitrogen

status of the leaf at the time of fruit bud differentiation. For instance, you may have a deficiency of N
at the critical state of fruit bud differentiation;
then you may have rain or fertilizer may be applied
and you change the composition of the leaf to a normal
value. If you try to relate the leaf composition, say
in August, against the yield of fruit, and the potential
yield of that fruit was influenced earlier in the season
when nitrogen was deficient, then you have a false
situation. You are trying to relate the composition
of the leaf in August to the yield of the fruit in the
following season; when it was the composition of the
leaf in July which probably determined the crop
potential.

I think when one is dealing with perennial crops,
it is rather important to relate the composition of
the leaf to some critical stage in either the initiation
or the development of the fruit.

# BIOCHEMICAL APPROACHES
# IN THE STUDY OF THE NUTRITIONAL STATUS OF PLANTS


Session Leader:   Frederick C. Steward

# FUNCTIONAL ASPECTS OF MINERAL NUTRIENTS IN USE FOR THE EVALUATION OF PLANT NUTRIENT REQUIREMENT

## Avigdor Bar-Akiva

*The Volcani Institute of Agricultural Research, Bet-Dagan, Israel*

ABSTRACT

Our knowledge of the exact role of each of the essential nutritional elements is, in many respects, still far from complete. The vast amount of information accumulated over the past decades paves the way for consideration of a functional approach, i.e. plant cell as a basis for the study of nutrient requirements of plants. Such an approach is a logical consequence of plant nutrition studies and first attempts at practical application are as old as first discoveries about the functions of these elements; nonetheless development of this approach is slow and its practical application is rare. Further, a wide gap exists between the possibilities offered by knowledge acquired in the fields of modern biochemistry and molecular biology and their agricultural use as far as mineral nutrition of plants is concerned. The reasons for this situation may be summarized as follows:

1) Lack of specificity of certain systems or metabolic products tested for diagnostic purposes;

2) The sporadic nature of these systems, which lack a common basis of thought and operation.

In attempting to overcome some of these diffi-

Avigdor BAR-AKIVA. Head Div. Citriculture, Volcani Inst. Agric. Res., Bet Dagon (Israel) since 1968. b. 1925 Budapest (Hungary); 1954 M.Sc. (Agr.) and 1963 Ph.D., Heb. Univ. Jerusalem-Rehovot; 1955-65 Scientist and 1965-67 Research Leader in Citrus Nutr., Div. Citriculture, Volcani Inst.

culties, some relatively new methods and ideas were
introduced, such as studying changes in enzyme activity
on isozyme level or using the inducible and adaptive
nature of enzymes for diagnostic purposes.  Studying
enzyme induction and, more generally, following the
course of reactivation of metabolic processes in the
plant cell by means of the resupplied deficient nutrients, may result in a new approach, offering a common
platform and technique for the use of rather diverse
metabolic pathways as indicators in the evaluation of
mineral nutrition requirement of plants.

---

The physiological roles of cations and anions in
the plant cell, which constitute the so-called essential mineral nutrients, have been the subject of continuous investigation for several decades.  The extensive literature in this field of research has recently
been reviewed by Hewitt [23], Nason and McElroy [28]
and Evans [21].  Precise information on the function
of the various elements, their interrelations, and
their impact on other growth factors, would undoubtedly have practical implications in plant growth, especially for the determination of nutrient requirements.  The advance in investigations of enzyme systems and the discovery of mineral nutrient involvement
in these systems gave an impetus to the study of these
relations.  Some of these studies, besides their
purely scientific interest, aim at a practical application of these systems for diagnostic purposes.

It is not intended here to cover the literature
devoted to this topic but to present a few milestones
indicating approaches, concepts and ideas for further
studies.  The reports dealing with the subject may be
classified into two groups, one which investigates
the activity of the enzyme system itself and the other
which deals with the metabolic products - substrates
and endproducts - accumulated or diminished due to
changes in the enzymatic conditions.

ENZYME ACTIVITIES AS INDICATORS OF MINERAL DEFICIENCIES

The earliest work which traces back to the beginning of this century is that of Bertrand [14]. He was probably the first to suggest that a micronutrient – manganese – is actively involved in a catalytic enzymatic system. Bailey and McHargue [3] were apparently among the forerunners in modern research to investigate the association between nutrient status of high plants and enzyme activity. This report was followed by many others [29, 43, 31, 2, 27, 34] who defined more clearly the specificity of association between enzymes and metals or nutrients. In spite of the increasing awareness of the intimate relationship between nutrient elements and enzyme systems, a practical application of this knowledge, for diagnostic and nutrient status evaluation was only attempted in a few cases. Brown and Hendricks [16] were probably the first to suggest in 1952 the use of enzyme activity for diagnostic purposes for copper and iron. This was followed by the proposal to use ribonuclease activity for the determination of zinc deficiency in orchard trees [25] and more recently the ascorbic acid oxidase as an index of available Cu [31]. The significance of these works lies in the fact that they propose enzyme activity assay as indicator for suboptimal values of nutrients before the appearance of the visible symptoms.

The measurement of peroxidase activity which decreases in the case of iron deficiency and increases in case of Mn deficiency has been proposed for distinguishing overlapping symptoms of these two deficiencies [4]. This test was successfully applied under field conditions in citrus [6] and peanuts [26]. To resolve apparently combined Fe, Mn and Zn deficiencies in citrus, besides the peroxidase test, the carbonic anhydrase test, a zinc metalloenzyme assay, is suggested as a specific indicator of zinc deficiency [9].

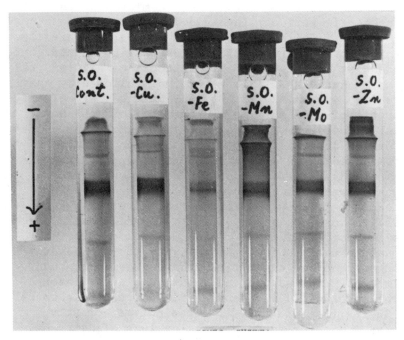

ANIONIC SYSTEM

PLATE 1.   Effect of micronutrient deficiencies in com-
           parison with control (cont.) on the anionic
           peroxidase enzyme patterns of Sour Orange
           leaf extracts.  The isoenzymes were separated
           by polyacrylamide gel electrophoresis [12].

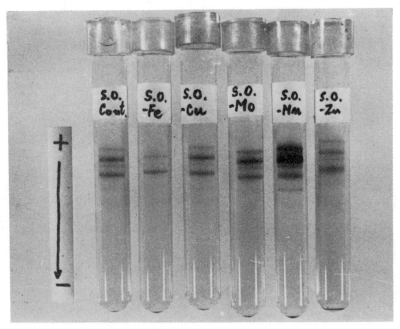

CATIONIC SYSTEM

PLATE 2.  Effect of micronutrient deficiencies in com-
          parison with control (cont.) on the cationic
          peroxidase isoenzyme patterns of Sour Orange
          leaf extracts.  The isoenzymes were separated
          by polyacrylamide gel electrophoresis [12].

The specific differences in peroxidase activity
between Mn and Fe deficient citrus leaf tissues can be
demonstrated even on the isoenzyme level.(Plate 1 and 2).

The results of the electrophoretic separation of
isoperoxidases in citrus leaves showing various deficiencies disclose at least four anionic and five
cationic isoenzymes.  The relative amount of the different isoperoxidases corresponds to the total activity values [4] as can be deduced from the outstand-
ing differences found between the Mn and Fe deficient
leaf extract (Plate 3).

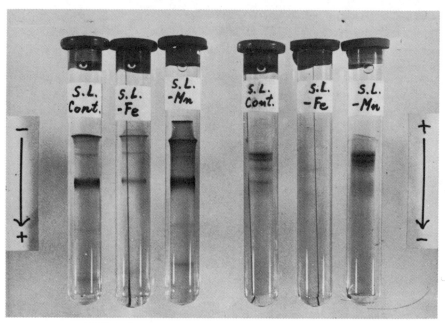

PLATE 3.  Differences in peroxidase isoenzyme
          patterns between iron and manganese
          deficient Sweet Lime leaf extract.

METABOLIC PRODUCTS AS INDICATORS OF DEFICIENCIES

   Suppression or acceleration of enzyme activities
due to deficiencies of mineral nutrients may bring
about either the accumulation of certain metabolic
products or their disappearance.  It has been assumed
in consequence that many of the visible deficiency
symptoms may be interpreted on the basis of upset metabolic conditions.  Thus Mo deficiency symptoms may, to
a certain extent, be interpreted as nitrate toxicity
due to the accumulation of nitrate caused by the absence
of Mo [22]; K deficiency symptoms as putrescine induced
toxicity [35, 18].  A great number of works describe
such an upset of metabolic patterns as a consequence
of mineral deficiencies.  The most extensively-treated
metabolic compounds are the amino acids, due to their
importance as intermediates in nitrogen metabolism, and
due to the attractability of their separation and detection following the development of chromatographic
methods [19].

   The effects of mineral nutrients on amino acid
composition of plants have been recently reviewed by
Hewitt [24] and Steward and Durzan [39].  The vast information accumulated on this subject is impressive, but
it also raises problems involved in the use of these
substances for the purpose of nutrient status evaluation.

   As can be deduced from the data, various nutrient
deficiencies affect the same amino acid level similarly,
but changes in environmental conditions do this in some
cases too.  Due to this lack of specificity, only a few
of the nitrogen intermediates can be useful in diagnosis
of nutritional problems.  Thus the interesting suggestion of Ozaki [30], concerning the application of the
asparagine accumulation test as an assay for the determination of nitrogen top dressing requirement in rice
fields, become questionable since the accumulation of
asparagine may be also caused by iron or zinc

## TABLE 1

Effect of Nutritional Deficiencies and Environmental Condition of
Free Amino Acids in Leaves of Various Plants

| Amino Acid | Plant | N | P | K | Ca | Mg | S | Fe | Mn | Zn | Cu | Mo | B | Gibber-elin | Short day | Long day | Chill-ing | Refer-ences |
|---|---|---|---|---|---|---|---|---|---|---|---|---|---|---|---|---|---|---|
| Argi-nine | Straw-berry | | | | + | + | | | | | | | | | + | | + | 15 |
| | Tomato | | | | | + | | + | + | + | + | + | | | | | | 32 |
| | Orange | | | | | + | | + | 0 | + | 0 | | | | | | | 41 |
| | Lemon | | | | | | | + | + | | | | | | | | | 5 |
| | Tobacco | | | | | | | + | + | + | + | + | + | | | | | 37 |
| | Peas | | | | | | | +± | | | | | | | | | | 20 |
| | Mentha | | | | | | + | | | | | | | | | + | | 38 |
| Glu-tamic Acid | Straw-berry | | | | − | − | | | | | | | | + | | + | + | 15 |
| | Tomato | | | | | | | + | + | + | + | − | | | | | | 32 |
| | Orange | | | | | − | | 0 | 0 | + | + | | | | | | | 41 |
| | Lemon | | | | | | | − | 0 | | | | | | | | | 5 |
| | Tobacco | | | | | | | − | − | − | − | | − | | | | | 37 |
| | Peas | | | | | | | + | | | | | | | | | | 2C |
| | Mentha | − | − | − | − | 0 | − | − | | | | | | | | + | | 38 |

+ Increase
− Decrease
0 No change

deficiency [32].  For the same reason it is doubtful
whether arginine content may be a reliable indicator
for assessing N status, or whether the determination
of total amount of amino N may be an indicator for N
status as was proposed by several Australian investigators [42, 13].

The accumulation of putrescine, which appears to
be specific as regards disturbances in potassium nutrition of plants [35] seems more promising for diagnostic purposes.  The increase of tryptophan content in
Mg deficient citrus leaves appears likewise specific to
this nutrient deficiency [42].  Among the non-nitrogen-
ous compounds it is worth mentioning the accumulation
of xylose which has been found specific for Mn deficiency in citrus leaves (Plate 4) [6].


BIOCHEMICAL APPROACH: FACTS AND IDEAS


It is a common feeling among plant nutrition
scholars, apparently shared even by those sceptical
of the biochemical approach, that in order to make a
successful interpretation of leaf analysis, a better
understanding of the basic physiological role of nutrients in the plant cell is essential.  A great deal
of this knowledge already exists, but no way has yet
been found to harness and use this basic knowledge in
the field and orchard.  One of the main reasons for
this is probably the lack of sufficient communication
between the plant physiologist and the biochemist on
the one hand, and the horticulturist on the other.
Apart from the psychological disinclination of the horticulturist to be concerned with "enzymes," and the
reluctance of the biochemist to consider the plant as
a whole and not only in purified enzyme form, there
exist fundamental problems impeding the development of
the biochemical approach as a practical tool.  These
problems may be summarized as follows:

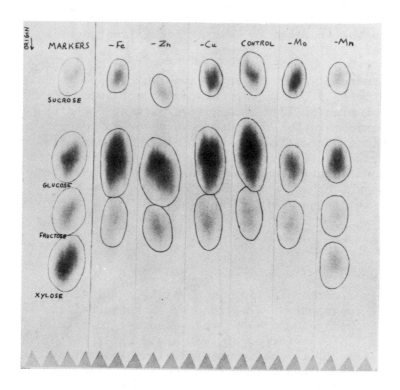

PLATE 4.   Chromatogram of sugars in the
           alcohol-water leaf extract from
           Eureka lemon grown in micronu-
           trient deficient cultures, showing
           xylose accumulation in the manga-
           nese deficient leaves [7].

1.  Lack of specificity of many metabolic products or enzyme systems for a specific element.

2.  Lack of a common basis for compounds and processes which have been suggested for these purposes.

Dealing with the first problem - a qualifying specificity test for each of the proposed indicators for the element in question is a prerequisite.  Table 2 illustrates such a specificity test for carbonic anhydrase activity as a measure of Zn deficiency.

The incorporation of sporadic trials with distinctive processes into a common conception and framework of action is a difficult task.  The conception and framework of action of the conventional foliar analysis are defined by the "standard curve" and the tentative tables of leaf standard values.  The shortcomings of this conception and the well known difficulties encountered in the use of standard values led to the search of improved indicators.  The establishment of standard values with biochemical indicators is not likely because of their diversity and their increased sensitivity of environmental factors.

A solution for both these problems may possibly be found in pursuing some of the ideas put forward by Steward [39].  He discusses the interaction between mineral nutrition and nitrogen metabolism or more specifically the effects of Mn upon the free amino acids and amides of the tomato plant [40].  These ideas are best presented by Steward in the following quotation [39]: "Incidentally, the best way to find the impact of the nutrient elements on the nitrogen metabolism is not merely to describe the nitrogen compounds which accumulate when the element in question is grossly deficient. A better method is to establish the symptoms of deficiency, at a reversible level and then study the changes that ensue with time when the deficient element is resupplied."

TABLE 2

Effect of Micronutrient Deficiencies on the Carbonic Anhydrase (A)
Activity and Zn Content of Three Varieties of Citrus Leaves (9)

| Treatment | Sour orange | | Eureka lemon | | Sweet lime | |
|---|---|---|---|---|---|---|
| | RCAA | Zn* | RCAA | Zn* | RCAA | Zn* |
| Control | 38.4±9.5 | 14.0 | 26.4±2.8 | 19.0 | 37.6±5.7 | 18.7 |
| -Zn | 11.5±1.7 | 6.0 | 7.2±1.3 | 13.4 | 9.0±3.6 | 11.8 |
| -Mn | 33.3±7.7 | 14.0 | 19.6±4.3 | 36.7 | 28.5±0.7 | 28.0 |
| -Fe | 42.2±2.3 | 16.0 | 32.6±4.7 | 30.9 | 40.5±4.0 | 44.0 |
| -Mo | 32.1±8.1 | 14.0 | 16.3±2.4 | -- | 38.6±5.5 | 44.0 |
| -Cu | 37.0±8.9 | 10.0 | 25.6±4.2 | 21.6 | 26.2±1.5 | 16.0 |

* ppm in dry material

RCAA = Relative Carbonic Anhydrase Activity (time in seconds
of uncatalyzed reaction - time of catalyzed reaction).

It appears that this approach may be applied in a
more generalized form for other metabolic products and
processes.  This approach can be adapted as a basis for
diagnosing nutrient needs, through measuring of so called
"reactivation processes" by means of the resupplied element.  These reactivation processes may be classified
into following groups:

1. *Reactivation or restoration of certain metabolic
   products.*

In case of Mn deficiency [40] in tomato, the resupplying of Mn (in the presence either of nitrate or
of ammonia) resulted in a rapid formation of dicarboxy-
lic acid and glutamine and a decrease of other amides.

2. *Restoration of certain metabolic processes.*

The involvement of Mn in the Hill reaction, or to
be more specific, in the process of 0 evaluation in
photosynthesis [17, 33], may be taken as a prototype
for the above process.  Brown et al. [17] succeeded in
obtaining a full recovery of the decreased Hill reaction
in the Mn deficient cultures of *Chlorella* through the
addition of Mn.  Preincubation of the algae for 30-60
minutes with Mn brought about Hill reaction capacity
increases by factors up to 15 and 20.  A greater state
of deficiency resulted in a higher percentage of recovery, which means that the rate of response to the Mn
supply is a function of the initial Mn nutrition status.
Possingham [33] working with separated chloroplasts of
Mn deficient tomato leaves did not attain recovery of
their normal oxygen evolution capacity in response to
Mn addition to the chloroplasts.  He demonstrated, however, that in solution cultures of intact tomato plants,
with suboptimal Mn nutrient conditions, the Hill activity depended on the Mn level in the substrate.  The
resupply of Mn in these cultures after 48 hours

increased the oxygen evolving capacity of the chloroplasts of these plants three times.  Our preliminary experiments with Mn solution infiltration into intact Mn
deficient lemon leaves also led to increased oxygen
evolution.  All these data indicate that the measure of
Hill reaction activity and its process of restoration
may potentially become a suitable indicator for Mn requirement of plants.


3.  *Reactivation or induction of enzyme activity.*

    Enzymes appear especially well fitted for "resto-
ration processes" due to their adaptive nature and the
important role mineral nutrients play in enzyme systems.  We have tested several enzyme systems for the
evaluation of nutrient requirement of citrus such as
nitrate reductase for N [11] and Mo [36], ascorbic oxidase for Cu [10], and peroxidase for Fe [8].  The
changes in peroxidase activity in iron deficient citrus
leaves, induced by the resupply of iron, were demonstrated even on the isoenzyme level by means of gel
electrophoresis techniques (Plates 4 and 5), showing
quantitative as well as qualitative differences.  One
of the four isoenzymes (A$_3$) was completely missing in
the iron deficient leaves, and in the case of cation
peroxidase isoenzymes, there was an almost total lack of
activity.  What appears significant is that the resupply of iron in detached, intact leaves, during 48
hours' incubation restored nearly fully the activity of
anionic and a great deal of cationic peroxidases without causing marked changes in the control.

    These results confirm previous findings [8] that a
close positive relationship exists between the iron
level in the substrate, the peroxidase activity of
citrus leaves, and the chlorophyll content of these
leaves.  The rate of restored or induced activity as a
result of resupplying of iron is negatively correlated
with the iron level in the substrate and the leaf
chlorophyll, thus producing a gap between initial and

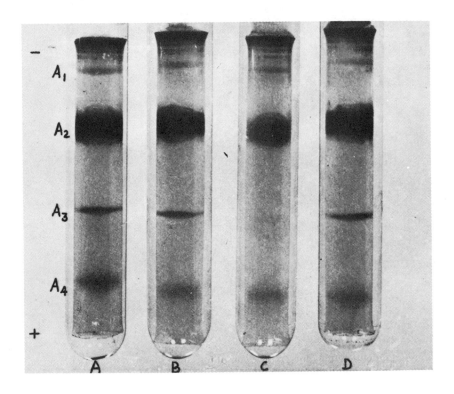

ANIONIC SYSTEM

PLATE 5.  Effect of iron solution infiltration into
          excised intact iron deficient and control
          leaves on the peroxidase isoenzyme pat-
          terms.  $A_1 - A_4$ anionic isoenzymes; A =
          control, water infiltrated, B = control,
          $FeSO_4$ infiltrated, C = iron deficient
          water infiltrated, D = iron deficient
          $FeSO_4$ infiltrated (12).

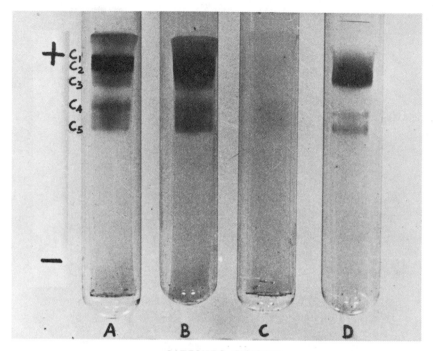

CATIONIC SYSTEM

PLATE 6.   Effect of iron solution infiltration into
           excised intact iron deficient and control
           leaves on the peroxidase isoenzyme pat-
           terns.  $C_1$ - $C_4$ cationic isoenzymes ; A =
           control, water infiltrated, B = control,
           $FeSO_4$ infiltrated, C = iron deficient
           water infiltrated, D = iron deficient
           $FeSO_4$ infiltrated (12).

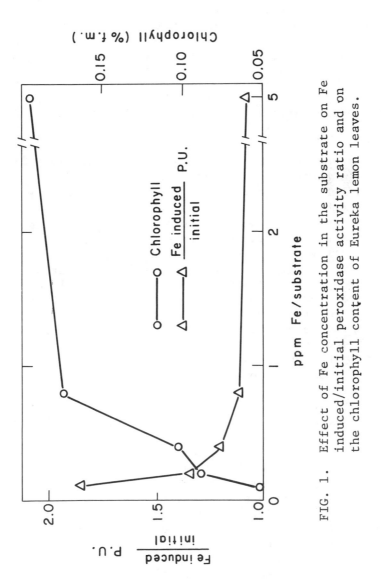

FIG. 1. Effect of Fe concentration in the substrate on Fe induced/initial peroxidase activity ratio and on the chlorophyll content of Eureka lemon leaves.

induced enzyme activity (Fig. 1). Consequently, the
rate of response to the enzyme induction, which is high
in the Fe deficient leaves and nearly nil in the normal
iron fed leaves, may serve as indicator for the iron requirement of the respective plants. Similar relationships, and the use of rate of response to enzyme induction were established for N [11], Mo [36] and Cu [10].


CONCLUDING REMARKS


     The conventional foliar analysis is widely ac-
cepted in the study of nutrition problems and as a
practical fertilizer guide in every day agriculture.
The contribution of foliar analysis in the last 3-4 decades to the development of modern agriculture was indeed remarkable. Yet, many researchers and extension
workers arrived at a conclusion that the benefits that
can be derived from foliar analysis for further development reached a certain point of exhaustion. It also became clear that the difficulties of foliar analysis stem
from its empirical nature and that future development
will depend on a better understanding of the physiological roles of the nutrients and of the metabolic processes
in which the nutrients are involved. This motivated a
search for new parameters for plant nutrient requirement evaluation, to be based on physiological processes,
aiming at obtaining more integrated data about the
present and potential activity of the elements in question.

     The development of new methods was heretofore
rather slow. The complexity of cell machinery, the
versatility of processes in which nutrient elements take
part and the sporadic nature of the few attempts at
practical application of such indicators, discouraged
the progress in this field.

     The new approach advocates the concept of "process

of reactivation" whereby the plant potential is measured
to respond to the supply of the specific element.  It is
hoped that this approach may become a turning point for
intensive explorations of the vast array of possibilities
offered by the cell machinery.

The aim of this communication is not to present a
certain enzyme system or a metabolic process which will
do well as indicator for nutrient requirement evaluation,
but to suggest a concept which may constitute a common
basis for different indicators proposed by the various
authors.

This concept may serve as a framework for tests already established, as a template for those to be elab-
orated and a stimulus for further endeavors in this
field.  A wide gap exists between the possibilities offered by knowledge gained in the fields of modern biochemistry and molecular biology and their practical
agricultural application as far as mineral nutrition is
concerned.  Possibly the main problem which modern agriculture faces is the exercise of more precise control
on the plant cell machinery for obtaining optimum growth,
and production under given environmental conditions.
The application of growth substances made a considerable
contribution in this direction.  No doubt, the intelli-
gent and less empiric use of nutrients based on metabolic processes may be an important step towards achiev-
ing this control.

SUMMARY

The knowledge of the exact role of each of the
essential nutritional elements is still far from complete and in some cases lacking.  The vast information
accumulated during the last 3-4 decades paves the way
towards a functional approach to this problem, i.e., the
examination of the nature and of the rate of action of

the element in the plant cell as a basis for the study
of nutrient requirement of plants. However, the development of this approach is slow and its practical application is rare. The apparent reasons for this are
a) lack of specificity of systems or metabolic products
tested for diagnostic purposes, and b) the sporadic
nature of these systems, lacking a common basis of approach and operation.

In attempting to overcome some of these difficulties the introduction of relatively new methods
and ideas was tried such as studying changes in
enzyme activity on isozyme level or using the in-
ductible and adaptive nature of enzymes for diagnostic
purposes. Enzyme induction and in a more generalized
form, the reactivation of metabolic processes in the
plant cell by means of the resupplied deficient nutrients, may offer a common platform and techniques
for the use of diverse metabolic indicators for the
evaluation of mineral nutrient requirements of plants.

REFERENCES

1.    AGARWALA, S.C. and SHARMA, C.P. (1961) The re-
      lation of iron supply to the tissue concen-
      tration of iron chlorophyll and catalase in
      barley plants grown in sand culture. *Physi-
      ologia Pl. 14*: 275-283.

2.    AGARWALA, S.C., SHARMA, C.P. and FAROOQ, S. (1965)
      Effect of iron supply on growth, chlorophyll,
      tissue iron and activity of certain enzymes in
      maize and radish. *Pl. Physiol. 40*: 493-499.

3.    BAILEY, L.F. and McHARGUE, J.S. (1944) Effects of
      boron, copper, manganese and zinc on the enzyme
      activity of tomato and alfalfa plants growing
      in a greenhouse. *Pl. Physiol. 19*: 105-116.

4.  BAR-AKIVA, A. (1961) Biochemical indications as a
    means of distinguishing between iron and manga-
    nese deficiency symptoms in citrus plants. *Nat-
    ure 190*: 647-648.

5.  ——————— (1964) Visible symptoms and chemical
    analysis vs. biochemical indicators as a means
    of diagnosing iron manganese deficiencies in cit-
    rus plants. *Plant Analysis and Fertilizer Prob-
    lems IV*: 9-24. Bould, C., Prevot, P. and Mag-
    ness, J.R. (eds.). Amer. Soc. Hort. Science,
    Humphrey Press Inc., Geneva, New York.

6.  BAR-AKIVA, A., KAPLAN, M. and LAVON, R. (1967) The
    use of a biochemical indicator for diagnosing
    micronutrient deficiencies of grapefruit trees
    under field conditions. *Agrochimica 11*: 285-
    288.

7.  BAR-AKIVA, A. and LAVON, R. (1967) Visible symp-
    toms and some metabolic patterns of micronutri-
    ent deficient Eureka lemon leaves. *Israel J.
    Agric. Res. 17*: 7-16.

8.  _____ (1968) Peroxidase ac-
    tivity as an indicator of the iron requirement
    of citrus trees. *Israel J. Agric. Res. 18*:
    144-153.

9.  _____ (1969) Carbonic anhy-
    drase activity as an indicator of zinc defic-
    iency in citrus leaves. *J. Hort. Sci. 44*: 359-
    362.

10. BAR-AKIVA, A., LAVON, R. and SAGIV, J. (1970) As-
    corbic acid oxidase activity as a measure of the
    copper nutrition requirement of citrus trees.
    *Agrochimica 14*: 47-54.

11. BAR-AKIVA, A. and STERNBAUM, J. (1965) Possible
    use of the nitrate reductase activity as a meas-
    ure on the nitrogen requirement of citrus trees.
    *Pl. Cell Physiol. 6*: 575-577.

12. BAR-AKIVA, A. and SAGIV, J. (1969)  Induced perox-
    idase isoenzyme patterns in citrus leaves.  *Ex-
    perientia 25*: 474-475.

13. BAXTER, P. (1965)  A simple and rapid test using
    the ninhydrin method, for the determination of
    the nitrogen status of fruit trees.  *J. Hort.
    Sci. 40*: 1-12.

14. BERTRAND, G. (1905)  Sur l'emploi favorable du
    manganèse comme engrais.  *C.R. Acad. Sci. (Paris)
    141*: 1255-1257.  (Cited in Stiles, W., Essential
    micro (trace) elements.  *Encyclopedia of Plant
    Physiology IV*: 558-614.  Springer Verlag, Berlin)

15. BOYNTON, D., YATSU, L. and KWONG, S.S. (1961)
    Some factors influencing the intermediary nitro-
    genous compounds in leaves of the strawberry
    plant.  (*Fragaria chiloensis* var. *amanassa),
    Plant Analysis and Fertilizer Problems*.  Reuther,
    W. (ed.).  American Inst. Biol. Sci., Washington,
    D.C.

16. BROWN, J.C. and HENDRICKS, S.B. (1952)  Enzymic
    activities indications of copper and iron defic-
    iencies in plant.  *Pl. Physiol. 27*: 651-660.

17. BROWN, T.E., EYSTER, H.C. and TANNER, H.A. (1958)
    Physiological effects of manganese deficiency.
    *Trace Elements*.  Lamb, C.A., Bentley, O.G. and
    Beattie, J.M. (eds.).  Academic Press Inc., New
    York.

18. COLEMAN, R.G. and RICHARDS, F.J. (1956)  Physio-
    logical studies in plant nutrient.  XVIII: Some
    aspects of nitrogen metabolism in barley and
    other plants in relation to potassium deficiency.
    *Ann. Bot. 20*: 393-409.

19. DENT, C.E., STEPKA, W. and STEWARD, F.C. (1947)
    Detection of the free amino acids of plant cells
    by partition chromatography.  *Nature 160*: 682.

20. DE KOCK, P.C. and MORRISON, R.I. (1958) The metabolism of chlorotic leaves: Amino acids. *Biochem. J. 70*: 266-272.

21. EVANS, H.J. and SORGER, G.J. (1966) Role of mineral elements with emphasis on the univalent cations. *A. Rev. Pl. Physiol. 17*: 47-76.

22. HEWITT, E.J. (1956) Symptoms of molybdenum deficiency in plants. *Soil Sci. 81*: 159-172.

23. _____ (1958) The role of mineral elements in the activity of plant enzyme systems. *Encyclopedia of Plant Physiology. IV*: 427-481. Rauhland, E. (ed.). Springer Verlag, Berlin.

24. ─────── (1963) The essential nutrient elements requirements and interactions in plants. *Plant Physiology: A Treatise. III*: 137-360. Steward, F.C. (ed.). Academic Press, New York.

25. KESSLER, B. (1961) Ribonuclease as a guide for the determination of zinc deficiency in orchard trees. *Plant Analysis and Fertilizer Problems.* 314-322. Reuther, W. (ed.). American Institute of Biological Sciences, Washington 6, D.C.

26. LACHOVER, D. and EBERCON, A. (1969) Iron deficiency problems in peanuts under irrigation. *Proc. Colloquium on the Transition from Extensive to Intensive Agriculture, Tel-Aviv, March 24-28.* International Potash Institute, Berne.

27. MULDER, E.G., BOXMA, R. and VAN VEEN, W.L. (1959) The effect of molybdenum and nitrogen deficiencies on nitrate reduction in plant tissues. *Pl. Soil 10*: 335-355.

28. NASON, A. and MC ELROY, W.D. (1963) Modes of action of the essential mineral elements. *Plant Physiology. A Treatise. III*: 451-536. Steward, F.C. (ed.). Academic Press, New York.

29.  NASON, A., OLDERWURTEL, H.A. and PROPT, L.M. (1952)
     Changes in oxidative enzyme constitution of to-
     mato leaves deficient in a micronutrient nutri-
     ents. *Arch. Biochim. Biophys. 38*: 1-13.

30.  OSAKI, K. (1961)  The detection of asparagine as
     a criterion for top dressing for rice in the
     field. *Plant Analysis and Fertilizer Problems*.
     323-325. Reuther, W. (ed.).  American Institute
     of Biological Sciences, Washington 6, D.C.

31.  PERUMAL, A. and BEATTIE, J.M. (1966)  Effect of
     different levels of copper on the activity of
     certain enzymes in leaves of apple. *Proc. Am.
     Soc. Hort. Sci. 88*: 41-47.

32.  POSSINGHAM, J.W. (1957)  The effect of mineral nu-
     trient on the content of free amino acids and
     amides in tomato plants.  I: A comparison of ef-
     fects of deficiencies of copper, zinc, manganese,
     iron and molybdenum. *Aust. J. Biol. Sci. 9*:
     539-551.

33.  POSSINGHAM, J.W. and SPENCER, D. (1961)  Manganese
     as a functional component of chloroplasts. *Aust.
     J. Biol. Sci. 15*: 58-68.

34.  QUINLAN, W.F. (1953)  The effect of zinc defic-
     iency on the aldolase activity in the leaves of
     oats and clover. *Biochem. J. 53*: 457-460.

35.  SAMISH, R.M. and HOFFMAN, M. (1966)  Free nitro-
     genous compounds as an indicator for the potas-
     sium nutrition status of fruit trees. *Proceed-
     ings of the XVII Int. Hort. Cong. 1*: Marshall,
     R.E. (ed.).  Michigan State University, East
     Lansing, Michigan.

36.  SHAKED, A. and BAR-AKIVA, A. (1967)  Nitrate re-
     ductase activity as an indication of molybdenum
     level and requirement of citrus plants. *Phyto-
     chemistry. 6*: 347-350.

37. STEINBERG, R.A. (1956) Metabolism of inorganic nitrogen by plants. *Inorganic Nitrogen Metabolism.* 153-158. McElroy, W.D. and Glass, B. (eds.). John Hopkins Univ. Press, Baltimore, Maryland.

38. STEWARD, F.C., CRANE, F., MILLAR, K., ZACHARIAS, R.M., RABSON, R. and MARGOLIS, D. (1959) Nutritional and environmental effects on the nitrogen metabolism of plants. *Utilization of Nitrogen and its Compounds by Plants. Symposia Soc. Exptl. Biol. 13:* 148-176.

39. STEWARD, F.C. and DURZAN, D.B. (1965) Metabolism of nitrogenous compounds. *Plant Physiology. A Treatise. IVa:* 379-686. Steward, F.C. (ed.). Academic Press, New York.

40. STEWARD, F.C. and MARGOLIS, D. (1962) The effects of manganese upon free amino acids and amides of the tomato plant. *Cont. Boyce Thompson Inst. Pl. Res. 21:* 393-410.

41. STEWART, J. (1962) The effect of minor element deficiencies on free amino acids in citrus leaves. *Proc. Amer. Soc. Hort. Sci. 81:* 244-249.

42. TAYLOR, B.K. and MAY, L.H. (1967) The nitrogen nutrition of the peach tree. II: Storage and mobilization of nitrogen in young trees. *Aust. J. Biol. Sci. 20:* 389-411.

43. WOOD, J.G. and SIBLY, D.M. (1952) Carbonic anhydrase activity in plants in relation to zinc content. *Aust. J. Scient. Res. B5:* 244-255.

*Questions to Dr. Bar-Akiva*

MALAVOLTA:  Has there been any work on zinc enzymes,
like carbonic anhydrase?

BAR-AKIVA: Carbonic anhydrase may serve as a good
indicator for zinc deficiencies, but it is a very
"stubborn" enzyme as far as its induction is concerned.
So this enzyme reflects fairly well the zinc level of
the substrate, but we can't get the enzyme inducted
with infiltrated zinc solution, and therefore we have
to renounce the use of the enzyme as a means of evaluating Zn requirement by means of the reactivation process. We are now trying various zinc enzymes, like
aldolase and others for diagnostic purposes, but still
haven't got the proper enzymes to work with.

CHRIST:Can you tell us whether the enzyme reactions
can be used with all plant species or do you have only
experience with citrus?

BAR-AKIVA: Most of this work has been done on citrus
plants, but some of the assays have been tested also
for other plants. For instance, the nitrate reductase
method for nitrogen has been tested also for some forage
crops such as corn and rye grass. We wanted to look into
plants which naturally accumulate nitrates, since citrus
leaves do not normally contain any nitrate.  With corn
and rye grass, the test has worked very nicely.

WEHRMANN:  Can you give us some information on the relationship between the total content of manganese and
the enzyme activity in your experiments?

BAR-AKIVA:  Lately we haven't done any enzyme work with
manganese. We have started to do some work with manganese
on the oxygen evolution in photosynthesis as a means of
manganese deficiency evaluation. Nitrogen, zinc and
copper have shown under controlled conditions,in the open

and in the greenhouse, nice correlation with the corresponding enzymes' activities.

WAISEL:  You showed us that the induced activity of the enzymes by infiltration is not on the same level as the control. Would you say that those plants suffered damage?

BAR-AKIVA:  I can't tell you whether they have suffered irreversible damage or not. With iron, we have carried out experiments and have shown that the enzyme recovery after infiltration is a function of time. The problem is, that it is just a little difficult to keep the infiltrated tissue in a healthy condition for a long time, without some deterioration, and therefore we do not try to prolong the incubation more than 48 hours. For practical purposes, it is important that we get a significant rate of increase as a response to the infiltration, as this is for us the indication for the requirement which is under consideration.

STEWARD:  I gathered that certain isozymes are more responsive than others. Where you've got several isozymes of the same enzyme, you obtained a big response from one to iron and not from the others. Have you got any ideas on this?

BAR-AKIVA:  I am not sure whether in our case one has been more responsive than the other. I am more inclined to think that this is a question of quantity. Since we do not do a quantitative estimation of these isozymes, I can't say this with certainty.

STEWARD:  I think it's organ specificity. If you did it on leaves you might get one isozyme and if you did it on tubers or the stem, flowers or fruit it might be another.

ROUTCHENKO:  Est-ce qu'il n'y a pas une difficulté que résulte du fait que plusieurs métaux contribuent à l'activation d'un système enzymatique?

BAR-AKIVA:  Well, the advantage of this method is that
you actually measure the response to the metal in question. If you measure iron deficiency and you infiltrate
iron, it does not interest me, as a horticulturist, too
much, whether it is a virus induced iron deficiency or
whether a toxic metal has induced iron deficiency by
means of an antagonistic effect. What is important in
this case, is that you get a response to the iron infiltration which should indicate potential response to
iron of the same plant under field conditions and this
is what is important from the horticultural point of view.

LAVEE:  I would like to comment on the specificity of
the plant with respect to isozymes, as you mentioned
before. We do know that isozymes have specific, distinct functions, and lately it has been shown that
certain isozymes might have two different enzyme
functions, such as IAA oxidase and a peroxidase, while
other isozymes of the same separation would not be
active, but for one of these two enzymes. So there seems
to be a very definite difference in function between the
different isozymes and there might be a very strong
speficity in this case.

# ENZYME CHANGES IN PLANTS FOLLOWING CHANGES IN THEIR MINERAL NUTRITION

## Roderick L. Bieleski

*Department of Scientific and Industrial Research, Auckland, New Zealand*

ABSTRACT

The duckweed *Spirodela oligorrhiza* has been used in studying how changes in mineral nutrition affect the biochemistry of a plant. 1. *Presence of urea*. Plants become nitrogen deficient, do not contain urease, and do not metabolize any urea when urea is supplied above pH 5. Below pH 4.5 the plants grow normally, contain urease and metabolize urea. 2. *Presence of nitrate*. In absence of nitrate, the plants do not contain nitrate reductase. When nitrate and $NH_4^+$ are supplied together, nitrate reductase is formed, but nitrate is not utilized until after all the $NH_4^+$ has first been metabolized. When nitrate alone is supplied, nitrate reductase is formed and nitrate is utilized for normal growth. 3. *Absence of phosphate*. In absence of phosphate in the medium, acid phosphatase activity in the plant increases 10 - 50 ×. Most of this increase is due to formation of one isozyme, apparently located in the outer cell membrane.

These examples illustrate ways in which the simple induction of an enzyme by a substrate can be complicated by other factors. Firstly, the substrate may not,

Roderick L. BIELESKI. Physiol. in Fruit Res., D.S.I.R., Auckland (New Zealand) since 1958. b. 1931 Auckland, New Zealand; 1955 M.Sc. (Bot.), Univ. New Zealand; 1959 Ph.D. (Pl. Physiol.), Univ. Sidney; 1960-61 and 1969-70 Res. Scholar, Univ. Calif., Los Angeles (U.S.A.).

under some conditions, induce the enzyme.  Secondly,
the induced enzyme may not function under some conditions.  Thirdly, the induced enzyme may apparently
duplicate the function of one already present in the
tissue, and its location within the cell may then have
significance.  The importance of enzyme induction
phenomena in mineral nutrition of plants will be discussed.

---------------

This paper reviews and discusses the implications
of some studies that have been carried out by several
workers in the Nutrition Section of the Plant Diseases
Division, D.S.I.R., New Zealand.  All the studies reviewed have used the small floating water plant, *Spirodela oligorrhiza,* as a model system for studying some
of the effects of different nutrient conditions on the
plant.  Although these studies have also included some
investigations on the chemical composition of the plant
(Bollard, 1966; Bieleski, 1968a; Cook, 1968; Ferguson
and Bollard, 1969), the present paper is confined to
reviewing various changes in enzyme pattern which have
been observed.  The system that has been used for these
studies has several advantages.  Spirodela plants can
be grown sterilely, avoiding the problems associated
with bacterial contamination; free of soil with all its
complexities; and in large numbers in a small area
under completely reproducible and controlled conditions.
The nutrient environment of the plants can be changed
at any given time from one composition to another.
Growing, treating and sampling the plants is thereby
simplified, and highly reproducible results have been
obtained.

The first aspect of the work to be reviewed concerns the utilization of urea by Spirodela.  Attempts
to grow Spirodela on urea as the sole nitrogen source

(Bollard, 1966) led initially to confusing results.
Sometimes the plants would grow rapidly and healthily;
whilst at other times there was no growth, and the
plants became chlorotic and nitrogen-deficient in appearance, despite the presence of ample urea in the
medium.  After several possible causes had been investigated, it was found that the pH of the culture medium
appeared to control the ability of Spirodela to grow on
urea (Bollard, Cook and Turner, 1968; Bollard and Cook,
1968).  If the pH was low, below pH 4.0, Spirodela was
able to grow readily; but if it was higher, above pH
5.0, the plants did not grow at all.  If a medium of
high pH, which was not supporting any growth, had its
pH lowered, either naturally through cation uptake by
the plants, or artificially by addition of acid, then
growth resumed within a couple of days.

Why was there no growth in the presence of urea
under certain conditions?  One possibility was that
urea might not be able to enter the tissue at the
higher pH's.  Instead, it was found that non-growing
plants at higher pH's contained a higher concentration
of urea than did the growing ones at lower pH's.  This
suggested that urea was not being metabolized.  It was
then shown that appearance of the enzyme urease (responsible for hydrolyzing urea to $NH_3 + CO_2$) in the
tissue was also susceptible to the pH of the medium.
Normally, Spirodela growing in the absence of urea, on
either $NH_4^+$ or $NO_3^-$ as sole nitrogen source, does not
contain detectable urease activity.  When the nutrients
are changed, so that urea is now presented as sole nitrogen source, the plant does not form any urease if
the medium pH is high; and so the plant is unable to
metabolize the urea and cannot grow.  However, when the
pH is low, there is an induction of urease by the urea,
considerable urease activity appears, and the plants
are then capable of metabolizing the urea and using it
for growth.

In this first example, therefore, growth depends
on the formation, presumably through an induction process, of an enzyme not normally present in the tissue;
but there is a complication in that this induction can
be controlled in turn by an apparently extraneous factor, the pH of the external culture medium.

The second example concerns the utilization of
nitrate by Spirodela (Ferguson and Bollard, 1969;
Ferguson, 1969).  Once again, neither of the two enzymes necessary for nitrate metabolism (nitrate reductase and nitrite reductase) can be detected in plants
growing on ammonium as sole nitrogen source.  But when
either nitrite or nitrate is supplied in the medium
there is an immediate induction of both enzymes, and
considerable nitrite reductase and nitrate reductase
appear in the tissue.  The ratio of nitrite reductase
to nitrate reductase depends on whether nitrite or
nitrate is the inducing metabolite: 24/1 for nitrite
inducing, and 9/1 for nitrate inducing.  It seems that
in this way, the plant prepares itself to keep the concentration of nitrite, toxic to living organisms, at a
low level in the tissue.  The induction of these two
enzymes occurs whenever nitrite or nitrate is present
in the medium - it is not prevented by high pH, nor by
the additional presence of ammonium in the medium.

Thus, whenever nitrite or nitrate is present, the
tissue rapidly acquires the potential ability to utilize these two compounds for growth.  This is unlike the
case with urea, where under some conditions the enzyme
was not induced, thus preventing any utilization of the
substrate.  Nonetheless, there is still a complication,
but this time, instead of at the stage of enzyme induction, at the stage of the substrate utilization itself.
If the ammonium ion is also present in the medium, no
nitrate is utilized until after all the ammonium has
been incorporated into the tissue, despite the presence
there of ample nitrate reductase.  That is to say, the
presence of one substrate is completely preventing the

utilization of a second, even though the necessary enzymes have been formed.  Once again, the effect is not
due to any failure of the nitrate to enter the tissue:
for although the presence of ammonium does slow down
the nitrate uptake to some degree, accumulation is by
no means prevented, so that this cannot explain the
complete inaction of the nitrate reductase.  In contrast, nitrite, the immediate product of nitrate reductase action, *can* be utilized in the presence of ammonium.  This suggests that the steps in nitrate utilization from nitrite onwards are fully functional, and
that the limit to nitrate reduction must be set by a
blocking of nitrate reductase itself.  The most likely
inhibitor of nitrate reductase would be expected to be
either ammonium itself, or some immediate product of
its metabolism.  There are at least four such compounds,
known to be present in much higher concentration when
tissues are grown in ammonium rather than in nitrate or
nitrite - ammonium itself, glutamine, asparagine and
arginine.  However, not one of these compounds, either
singly or together, is an effective inhibitor of nitrate
reductase *in vitro*.  Nor does the presence of glutamine
or asparagine in the medium prevent nitrate utilization
*in vivo* (Ferguson, 1970).  An untested possibility is
that the relative levels of the reduced and oxidized
forms of the nicotine adenine nucleotides may be the
controlling factor.

In this second example, therefore, utilization of
nitrate depends on the formation of an enzyme not present in the tissue when nitrate is absent; but there is
a complication in that the action of the induced enzyme
can be completely controlled in turn by some other (unknown) factor.

The last part concerns the response of Spirodela
to phosphorus deficiency - that is, to removal of a nutrient rather than to presentation of a new one (Reid,
1968; Reid and Bieleski, 1970a, 1970b; Bieleski, 1970).
The enzyme being studied, acid phosphatase, hydrolyzes

phosphate esters to inorganic phosphate plus the corresponding alcohol: and inorganic phosphate, the product
of the reaction, is the nutrient which has been removed
from the medium.  The response of Spirodela to phosphorus deficiency is to increase its phosphatase activity 20- to 50-fold.  Superficially, this response
seems different to the two already described, in that
the enzyme which is being formed as a result of the nutrient change is also present to a smaller degree in
the normal tissue.  But is it exactly the same enzyme
which is being formed?  Apparently not, for when enzyme
extracts of normal and phosphorus-deficient plants are
studied by acrylamide gel electrophoresis, it is found
that much of the increase in phosphatase activity is
due to the appearance of two completely new phosphatase
isoenzymes in the tissue (Reid, 1968; Reid and Bieleski,
1970a).  Thus, though there is a "constitutive" phosphatase present normally, there can also be an "inducible" phosphatase that appears under the stimulus of
phosphorus deficiency.

At first sight, it would seem that the new phosphatase isoenzymes are produced, as part of an internal
salvage mechanism, to break down the less-important
phosphorus compounds, in order to provide inorganic
phosphate for synthesis of the more important ones.
However, there are difficulties in the way of this interpretation.  The amount of phosphatase present is
very high - sufficient to completely hydrolyze all the
phosphate esters contained in the tissue in less than
30 seconds.  Despite this, phosphate-deficient plants
show an almost unaltered phosphate ester pattern; and
the turnover rates of the various phosphate esters are
not greatly affected (Bieleski, 1968b).  This sharp
contrast between an extremely high enzyme activity, and
no sign of any enzyme action within the tissue, is difficult to explain on an ordinary control mechanism.  It
seems rather that the phosphatase must be segregated
from the internal phosphate ester pools in some way.
Selective staining procedures, which reveal the sites

where phosphatase is present in a plant or animal tissue, have been used, and these show that most of the phosphatase in the phosphorus-deficient tissue is localized in the lower epidermis and in the roots of the frond (Reid, 1968). Within the cells, most of the activity appeared to be located in, or near to, the cell wall.

Another type of experiment was then used to learn more about the localization of the phosphatase, and to show that much of the phosphatase of the deficient tissue was localized in the outer cell membrane, or in the cell wall itself (Bieleski, 1970). Tissue homogenates on the one hand and whole plants on the other were supplied with a suitable phosphatase substrate (either p-nitrophenylphosphate or glucose-1-phosphate) and the relative rates of hydrolysis were then compared. Results from one such experiment are shown in Table 1.

TABLE 1.  Phosphatase Activity in Homogenates and Whole Plants of Normal and Phosphorus-Deficient Spirodela

| Tissue | Measured Apparent Phosphatase Activity of | | Ratio of Activity |
| | Intact Tissue | Homogenate | Intact Tissue/ Homogenate |
| --- | --- | --- | --- |
| Normal | 60 | 1580 | 0.04 |
| Minus-P | 2920 | 9180 | 0.32 |

Substrate: 1.0 mM glucose-1-phosphate ($P^{32}$) + 1.0 mM $KH_2PO_4$. Phosphatase activity measured as amount of inorganic phosphate ($P^{32}$) activity released to external solution of whole tissues, or to solution of homogenate. Aliquots of solutions taken at various times, and inorganic phosphate and glucose-1-phosphate separated by thin-layer chromatography, then radioactivity of inorganic phosphate measured.

In the homogenates, the entire tissue content of phosphatase is accessible to the phosphatase substrate: in
the whole tissues, however, that part of the cell which
lies inside the cell membrane is virtually inaccessible
to the substrate, since neither p-nitrophenylphosphate
nor glucose-1-phosphate can readily pass through the
cell membrane.  Consequently, in the intact tissue, the
only phosphatase which can act on the substrate is that
phosphatase which lies outside the impermeable outer
cell membrane.  Thus the ratio of the two hydrolysis
rates (intact tissue rate/homogenate rate) indicates
the fraction of the total cell phosphatase which is located externally, on or outside the cell membrane.  In
normal tissues, only 4-10% (in different experiments
with either substrate) of the phosphatase behaved as if
it had an external location, whereas in phosphorus-
deficient tissue, 32-48% behaved in this way (Table 1).
It was also shown that hydrolysis of substrate by intact plants was not caused by the substrate diffusing
into the cell, being hydrolyzed there, and the products
then leaking out; since the measured leakage rate of
phosphate from the tissue was only 1% of the observed
hydrolysis rate.  Also, when glucose-1-phosphate ($P^{32}$)
was used as the substrate, it was found that the product of hydrolysis, inorganic phosphate ($P^{32}$), appeared
in the external solution of intact tissue long before
it appeared in the tissue itself (through subsequent
accumulation of the inorganic phosphate).

Thus it seems that the marked increase in phosphatase activity during phosphorus deficiency comes about
through the induction (more accurately, derepression)
of a new isoenzyme, which has a specific location,
either in the outer surface of the cell membrane, or
outside the membrane in the cell wall.  The most likely
role of such an enzyme is to hydrolyze any esters that
have been released into the external medium by dying
fronds or cells nearby, so that the surviving cells can
take up the released inorganic phosphate.

In this third example, therefore, it appears that a response to nutrient change may involve the formation of a new enzyme which apparently duplicates the properties of an enzyme already existing in the tissue; and that the actual site of location of this new enzyme in the cell may have a part to play in modifying or controlling its action, and thereby directing its function.

Each of these three examples illustrates a different way in which a change in the nutrient conditions of a plant can lead to marked changes in the enzyme complement of the tissue. In the first case, it was shown that under some conditions the inducer would not induce the enzyme to form; but that if the enzyme was formed, it acted. In the second case, the inducer caused the enzyme to form, but the presence of other tissue metabolites was able to prevent the enzyme from acting. In the third case, the newly formed enzyme, though apparently just increasing an activity already present in the tissue, proved on detailed analysis to be a separate enzyme, with a separate and segregated location in the tissue of the plant. It is hoped that these examples serve to show that enzyme induction phenomena are important in control of plant processes, but that the induction phenomena themselves tell us only one small part of the story, since other control processes can intervene as well.


REFERENCES


1.    BIELESKI, R.L. (1968a)  Levels of phosphate esters
         in Spirodela. *Plant Physiol. 43*: 1297-1308.

2.    _____ (1968b)  Effects of phosphorus de-
         ficiency on levels of phosphorus compounds in
         Spirodela. *Plant Physiol. 43*: 1309-1316.

3.   _____ In Preparation. The external lo-
     cation of an induced plant phosphatase.

4.   BOLLARD, E.G. (1966)  A comparative study of the
     ability of organic nitrogenous compounds to
     serve as sole sources of nitrogen for the
     growth of plants. *Plant Soil 25*: 153-166.

5.   BOLLARD, E.G. and COOK, A.R. (1968)  Regulation of
     urease in a higher plant. *Life Sci.* 7: 1091-
     1094.

6.   BOLLARD, E.G., COOK, A.R. and TURNER, N.A. (1968)
     Urea as sole source of nitrogen for plant
     growth.  I: The development of urease activity
     in *Spirodela oligorrhiza*. *Planta 83*: 1-12.

7.   COOK, A.R. (1968)  Urea as sole source of nitrogen
     for plant growth.  II: Urease and the metab-
     olism of urea in *Spirodela oligorrhiza*. *Planta
     83*: 13-19.

8.   FERGUSON, A.R. (1969)  Nitrogen metabolism of
     *Spirodela oligorrhiza*.  II: Control of the en-
     zymes of nitrate assimilation. *Planta 88*:
     353-363.

9.   _____ (1970) Nitrogen metabolism of
     *Spirodela oligorrhiza*.  III:.Amino acids and
     the utilization of nitrate. *Planta 90*: 365-369.

10.  FERGUSON, A.R. and BOLLARD, E.G. (1969)  Nitrogen
     metabolism of *Spirodela oligorrhiza*.  I: Util-
     ization of ammonium, nitrate and nitrite.
     *Planta 88*: 344-352.

11.  REID, M.S. (1968)  Response of Spirodela to phos-
     phorus deficiency.  Ph.D. Thesis, University of
     Auckland (N.Z.).

12.  REID, M.S. and BIELESKI, R.L. (1970a)  Changes in
     phosphatase activity in phosphorus-deficient
     Spirodela.  *Planta*, In Press.

13.  _____ (1970b)  Response of
     *Spirodela oligorrhiza* to phosphorus deficiency.
     *Plant Physiol. 45*: In Press.

# SOME EFFECTS OF MINERAL NUTRIENT DEFICIENCIES ON THE CHLOROPLASTS OF HIGHER PLANTS

John V. Possingham

*CSIRO, Division of Horticultural Research, Glen Osmond, South Australia, Australia*

ABSTRACT

The rates of photosynthesis (carbon dioxide fixation in the light expressed on a per unit chlorophyll basis) of mineral-deficient spinach plants has been measured using infrared gas analysis. All nutrient deficiencies except those of iron and molybdenum depressed photosynthesis on a per unit chlorophyll basis with deficiencies of manganese, copper phosphorus and potassium bringing about marked depressions. These results are in close agreement with earlier data in which the photochemical activities (as measured by Hill reaction rates) of chloroplasts isolated from the leaves of mineral deficient tomato plants were measured.

The nitrogen and chlorophyll content of chloroplasts from spinach plants has been shown to vary with leaf age and with the nutritional status of plants. Chloroplasts from manganese- and iron-deficient plants contain less chlorophyll but similar amounts of nitrogen to chloroplasts from full nutrient plants. Chloroplasts from sulphur and potassium deficient plants contain lower amounts of both chlorophyll and nitrogen than control plants.

John V. POSSINGHAM. Chief Div. Hort. Res., CSIRO, Adelaide, S. Australia, since 1962. b. 1929 Australia; 1953 B.Sc. and 1955 M.Sc., Univ. Adelaide; 1957 Ph.D., Univ. Oxford; 1952 Res. Scientist, CSIRO Div. Plant Indus., Canberra, Australia.

The numbers of chloroplasts per cell in the leaves of spinach plants has also been shown to vary with leaf age and with nutritional status. The cells of young iron deficient leaves have a greatly reduced number of chloroplasts per cell relative to the numbers present in the cells of control leaves of similar age. By contrast manganese deficiency does not greatly reduce chloroplast number per cell.

INTRODUCTION

In higher plants deficiencies of most of the essential mineral elements cause visual symptoms of abnormality in the leaves such as yellowing and chlorosis (Wallace, 1951).

This paper reviews the results of a series of investigations which provide information about the ways in which nutrient element deficiencies affect the nature and function of the photosynthetic apparatus of higher plants. Studies describing the effects of nutrient deficiencies on the carbon assimilation of intact plants and on the photosynthetic reactions of isolated chloroplasts and observations involving the light and the electron microscope of the leaf cells and chloroplasts of mineral deficient plants are included. In addition data are provided for spinach on the effect of mineral deficiencies on the numbers of cells per unit area, on the numbers of chloroplasts per leaf cell and on the chlorophyll and nitrogen contents of isolated chloroplasts.

CARBON ASSIMILATION RATES OF WHOLE PLANTS

The rates of photosynthesis (carbon dioxide fixed in the light expressed on a per unit chlorophyll or per

TABLE 1.  Effect of Nutrient Deficiencies on the Photosynthesis of Spinach Plants

| Deficiency | Photosynthesis | |
|---|---|---|
| Control | 46.0* | 135.7** |
| -N | 9.9 | 75.2 |
| -P | 17.3 | 55.1 |
| -K | 32.7 | 56.1 |
| -Mg | 21.6 | 94.7 |
| -S | 13.7 | 73.2 |
| -Ca | 43.0 | 100.8 |
| -Fe | 26.6 | 124.5 |
| -Mn | 11.5 | 50.4 |
| -Cu | 24.2 | 61.8 |
| -B | 36.8 | 78.1 |
| -Zn | 20.2 | 79.3 |
| -Mo | 23.2 | 134.4 |

\* $\mu$g $CO_2$ fixed/min./g. f. wt.

\*\* $\mu$g $CO_2$ fixed/min./mg. chlorophyll

unit fresh weight basis) for whole spinach plants have been measured using infrared gas analysis techniques (Bottrill, Possingham and Kriedemann, 1970). Measurements were made when the plants displayed clear visual symptoms of deficiency relative to control plants.

Plants deficient in all the nutrient elements except iron and molybdenum had depressed photosynthesis

when chlorophyll was the basis of calculation, with the
depressions in manganese, copper, phosphorus and potassium deficient plants being the greatest. Alterna-
tively when photosynthesis was calculated on a fresh
weight basis calcium was the only deficiency which had
no effect. These results show that virtually all mineral deficiencies affect carbon assimilation in spinach
and are generally similar to those reported for green
algae (Pirson, 1955) and for a range of other higher
plants (Bouma, 1967; Gerretson, 1949; Ruszkowska, 1960).

## PHOTOSYNTHETIC REACTIONS OF ISOLATED CHLOROPLASTS

The Hill reaction activity of chloroplasts isolated from the leaves of mineral deficient tomato
plants has been measured (Spencer and Possingham, 1960).
In these experiments Hill reaction activity was measured
using the dye O-chlorophenol-2,6-dichloroindophenol and
the results were expressed on a per unit chlorophyll
basis. All deficiencies except that of iron resulted
in chloroplasts with reduced activities.

In most cases the individual deficiencies inhibited
Hill reaction activity and carbon assimilation rates in
a similar manner. In no case was the activity of chloroplasts from deficient plants enhanced by the addition
of the deficient element to the chloroplast suspensions.
By contrast a rapid *in vivo* recovery of photochemical
activity occurs when manganese is supplied to manganese
deficient tomato plants (Possingham and Spencer, 1962),
while Bouma (1967) has shown that the photosynthesis of
clover plants deficient in sulfur or phosphorus recovers
rapidly when the limiting nutrients are supplied.

Manganese deficiency which markedly depresses both
overall photosynthesis and chloroplast photochemical
reactions also affects photophosphorylation (Spencer
and Possingham, 1961). Plastids from manganese defi-

TABLE 2. Nutrient Deficiency Effects on Leaf Anatomy

| Nutrient treatment | Leaf no.** | Leaf area ($cm^2$) | Cells $\times 10^3$ per disc* | Chloroplasts per cell | Chloroplasts $\times 10^5$ per disc* |
|---|---|---|---|---|---|
| Control† | 1 and 2 | 14 | 22 | 436 | 102 |
| | 3 and 4 | 42 | 46 | 387 | 189 |
| | 5 and 6 | 47 | 48 | 184 | 94 |
| −Fe | 1 and 2 | 5 | 37 | 273 | 106 |
| | 3 and 4 | 5 | 80 | 52 | 42 |
| | 5 and 6 | -- | -- | -- | -- |
| −Mn | 1 and 2 | 5 | 24 | 380 | 101 |
| | 3 and 4 | 11 | 82 | 155 | 130 |
| | 5 and 6 | 10 | 81 | 157 | 140 |
| Control†† | 1 and 2 | 22 | 22 | 557 | 120 |
| | 3 and 4 | 41 | 40 | 426 | 168 |
| | 5 and 6 | 28 | 100 | 125 | 125 |
| −N₂ | 1 and 2 | 4 | 24 | 303 | 70 |
| | 3 and 4 | 9 | 58 | 171 | 97 |
| | 5 and 6 | 4 | 219 | 64 | 136 |

\* Discs 5 mm in diameter      † Control −Fe and −Mn plants 17 days old
\*\* Leaves number from the base of the stem    †† Control and −N₂ plants 18 days old

cient plants have a reduced capacity for photophosphorylation reactions which involve oxygen evolution.  This result supports the suggestion first made
by Kessler (1955) that manganese is a component of the
oxygen evolving system in photosyntheses.  In both
higher plants and algae it has been shown that this
element is a bound constituent of the photosynthetic
apparatus (Possingham and Spencer, 1962; Cheniae and
Martin, 1966; Homann, 1967), and further work has
located this element as a functional component of
Photosystem II (Anderson and Thorne, 1968; Cheniae
and Martin, 1968).


## LEAF ANATOMY AND CHLOROPLAST ULTRASTRUCTURAL EFFECTS


     Deficiencies of all the essential nutrient el-
ements lead to pronounced reductions in the leaf area
of plants.  However few investigations have been made
of the influence of mineral deficiencies on cell size
and cell number in leaves.  In a recent series of
experiments we examined the effect of deficiencies of
iron, manganese and nitrogen on the leaf anatomy of
spinach plants (Possingham and Saurer, unpublished).
The techniques we used for measuring cell number per
unit area of leaf and chloroplast number per cell
were as described earlier (Possingham and Saurer, 1969;
Saurer and Possingham, 1970).

     The combined number of palisade and mesophyll
cells per 5 mm disc was increased in plants deficient
in each of these three elements suggesting that cell
size was reduced.  By contrast the average number of
chloroplasts per cell was lower in the deficient
leaves, the effects of iron and nitrogen deficiencies
being greater than that of manganese deficiency.  The
two effects in combination were such that chloroplast
number per disc was depressed in the upper leaves by

iron deficiency, and in the lower leaves by nitrogen
deficiency but changed very little with manganese
deficiency.

The effects that mineral deficiencies have on the
fine structure of higher plant chloroplasts have been
examined by a number of workers (Thomson and Weier,
1962; Vesk, Possingham and Mercer, 1966). Characteristic alterations of chloroplasts fine structure occur
in the case of iron and manganese deficiency (Bogorad
et al, 1959; Possingham, Vesk and Nittim, 1964),
whereas the ultrastructural changes associated with
other nutrient deficiencies vary considerably with
species, leaf age and growing conditions. The effect
of manganese deficiency on the ultrastructure of
chloroplasts has been examined intensively and evidence is available which suggests that this element
is a structural component of lamella stacks of
chloroplasts (Teichler-Zallen, 1969). Other elements
such as iron and copper are known chloroplast constituents, being components of ferridoxin and plastocyanin
respectively, but no convincing evidence is available
to suggest they play a structural role in chloroplasts.


COMPOSITION OF ISOLATED CHLOROPLASTS


We have examined the effect of a number of nutrient deficiencies on the chlorophyll and nitrogen content of spinach chloroplasts isolated under conditions
designed to prevent leaching losses (Bottrill and
Possingham, 1969a and b). Chloroplasts from manganese
and iron deficient plants contain less chlorophyll but
similar amounts of nitrogen to chloroplasts from
plants grown on full nutrient. Chloroplasts from
sulfur and potassium deficient plants contain lower
amounts of both chlorophyll and nitrogen than control.

CONCLUSIONS


This brief summary indicates that deficiencies of
virtually all the essential nutrient elements affect
the rate of photosynthesis and alter the photosynthetic
apparatus of higher plants.  In the case of the element, manganese, evidence has accumulated which
closely defines the enzymatic role that this element
plays in photosynthesis and, additional evidence indicates that this element has a structural role in
chloroplasts.  Although many other nutrient elements
are known chloroplast constituents, and in their absence chloroplasts have abnormal composition structure and function, there is as yet no evidence that
they play a direct role either enzymatically or struc-
turally on chloroplasts.


REFERENCES

1.    ANDERSON, JAN, M. and THORNE, S.W. (1968)  The
         fluorescence properties of manganese deficient
         spinach chloroplasts.  *Biochim. and Biophys.*
         *Acta 162*: 122-134.


2.    BOGORAD, L., PIRES, G., SWIFT, H. and MCILRATH,
         W.J. (1959)  *Brookhaven Symposium in Biology No.*
         *11*: 132-137.


3.    BOTTRILL, D.E. and POSSINGHAM, J.V. (1969a)  Iso-
         lation procedures affecting the retention of
         water soluble nitrogen by spinach chloroplasts
         in aqueous media.  *Biochim. and Biophys. Acta*
         *189*: 74-79.


4.    ———————————————————— (1969b)  The
         effect of mineral deficiencies and leaf age on
         the nitrogen and chlorophyll content of spinach
         chloroplasts.  *Biochim. and Biophys. Acta 189*:
         80-84.

5.  BOTTRILL, D.E., POSSINGHAM, J.V. and KRIEDEMANN, P.E. (1970) The effect of nutrient deficiencies on photosynthesis and respiration in spinach. *Plant and Soil 32*: 424-438.

6.  BOUMA, D. (1967) Growth changes of subterranean clover during recovery from phosphorus sulphur stresses. *Aust. J. Biol. Sci. 20*: 51-66.

7.  CHENIAE, G.M. and MARTIN, I.F. (1966) Studies on the function of manganese in photosynthesis. *Brookhaven Symposium in Biology No. 19*: 406-417.

8.  ———————————————————— (1969) Photo reactivation of manganese catalyst in photosynthetic oxygen evolution. *Plant Physiol. 44*: 351-360.

9.  GERRETSON, F.C. (1949) Manganese in relation to photosynthesis. *Plant and Soil 1*: 346-358.

10. HOMANN, P.H. (1967) Studies on the manganese of the chloroplast. *Plant Physiol. 42*: 997-1007.

11. KESSLER, E. (1955) On the role of manganese in the oxygen evolving system of photosynthesis. *Arch. Biochem. Biophys. 59*: 527-529.

12. PIRSON, A. (1955) Functional aspects in mineral nutrition of green plants. *Ann. Rev. Plant Physiol. 6*: 71-114.

13. POSSINGHAM, J.V. and SPENCER, D. (1962) Manganese as a functional component of chloroplasts. *Aust. J. Biol. Sci. 15*: 58-68.

14. POSSINGHAM, J.V., VESK, M. and MERCER, F.V. (1964) The fine structure of leaf cells of manganese deficient spinach. *J. Ultra. Struct. Res. 11*: 68-83.

15.  POSSINGHAM, J.V. and SAURER, W. (1969)  Changes in
     chloroplast number per cell during leaf develop-
     ment in spinach.  *Planta (Berl.) 86*: 186-194.

16.  RUSZKOWSKA, M. (1960)  Some experiments on the
     physiological role of manganese in tomato plants.
     *Acta Soc. Bot. Polon. 29*: 553-579.

17.  SAURER, W. and POSSINGHAM, J.V. (1970)  Studies on
     the growth of spinach leaves.  *J. Expt. Bot. 21*:
     151-158.

18.  SPENCER, D. and POSSINGHAM, J.V. (1960)  The effect
     of nutrient deficiencies on the Hill reaction of
     isolated chloroplasts from tomato.  *Aust. J. Biol.
     Sci. 13*: 441-455.

19.  ————————————————————————  (1961)  The effect
     of manganese deficiency on photophosphorylation
     and oxygen-evolving sequence in spinach chloro-
     plasts.  *Biochim. Biophys. Acta. 52*: 379-381.

20.  TEICHLER-ZALLEN, DORIS (1969)  The effect of man-
     ganese on chloroplast structure and photosyn-
     thetic ability of chlamydomonas reinhardi.
     *Plant Physiol. 44*: 701-710.

21.  THOMSON, W.W. and WEIER, T.E. (1962)  The fine
     structure of chloroplasts from mineral deficient
     leaves of phaseolus vulgaris.  *Amer. J. Bot.
     49*: 1047-1055.

22.  VESK, M., POSSINGHAM, J.V. and MERCER, F.V. (1966)
     The effect of mineral nutrient deficiencies on
     the structure of the leaf cells of tomato,
     spinach and maize.  *Aust. J. Bot. 14*: 1-18.

23.  WALLACE, T. (1951)  The diagnosis of mineral de-
     ficiencies in plants by visual symptoms.  H.M.
     Stationery Office, London.

*Questions to Dr. Possingham*

DE WIT:  First I want to know at what light intensity
did you work at, second, were you working at a normal
$CO_2$ level or a very high one and the third - did you
do any experiments or any measurements where the
deficiency was not very severe?

POSSINGHAM:  We didn't particularly look at the very
early incipient stages, this was a very crude sort
of approach to sorting out those deficiencies that
would have a big effect.

The intensity of the light was 2000 foot candles
and we did allow for a light saturation period. There
was an air recirculating system at normal levels of
300 parts per million $CO_2$.

# THE ROLE OF UNIVALENT CATIONS
# IN ENZYME SYSTEMS

Richard W. Wilson

*Dept. of Botany, University of Texas, Austin, Texas, U.S.A.*

## ABSTRACT

Several enzymes from a range of metabolically different pathways show a stimulation by an absolute requirement for a univalent cation for maximum activity. The requirement in most instances is met by $K^+$. Generally, $Rb^+$ and $NH_4^+$ may substitute effectively; $Na^+$ is a weak substitute and $Li^+$ is a completely ineffective substitute for $K^+$. Of the enzymes which has been shown to be activated by univalent cations, pyruvate kinase has been studied most extensively. Kinetic studies with pyruvate kinase have shown that $K^+$ affects the $V_{max}$ but not the $K_m$ for substrates. Similar conclusions have been reported with tryptophanase and aceto-thiokinase. These results may be best explained by assuming that $K^+$ influence the conformation of the enzyme molecule. Support for this conclusion has come from immuno-electrophoretic studies of the enzyme in different salt environments as well as ultraviolet difference spectra of the enzyme in high salt concentrations. These studies support the conclusion that the apoenzyme itself is affected by the univalent cation environment. It has been concluded

Richard W. WILSON. Asst. Prof. Bot., Univ. Texas, since 1968. b. 1939 Madison (Wisconsin, U.S.A.); 1961 B.A., Carleton Coll.; 1964 M.Sc. (Pl. Pathol.), N. Carolina State Univ.; 1967 Ph.D. (Pl. Physiol.), Oregan State Univ.; 1967 Res. Fellow, Dept. Bot., Univ. Illinois.

that a major function of $K^+$ in living systems may involve its role as an enzyme activator and that with at
least some enzymes this probably involves an altera-
tion of the protein structure.

---

Almost all forms of both plant and animal life
are dependent on the presence of potassium for normal
growth and development. In plants potassium constitutes the most abundant mineral and fifth most
abundant element exceeded in amounts only by the basic
biological elements of carbon, hydrogen, oxygen and
nitrogen (Epstein, 1965). While the essentiality of
potassium in living systems is well established, its
function in the metabolism of a living cell has re-
mained obscure. It has been suggested that the ion
participates in maintaining cell organization, cell
membrane permeability (Sutcliffe, 1962; Nason and
McElroy, 1963) and in the regulation of osmotic
pressure (Sutcliffe, 1962). While these roles may be
implicated in a general way in the function of
potassium, they fail to account for certain basic
aspects of the potassium requirement, including the
specificity for potassium in living systems and the
characteristic metabolic effects associated with the
deficiency of potassium.

An alternative explanation for the role of
potassium, which accounts for many observed phenomena
associated with its occurence in living cells, concerns its capacity of function as an activator of
enzyme systems. A summary of the different enzymes
which are stimulated by or have an absolute requirement for potassium has recently been reported (Evans
and Sorger, 1966). The list includes enzymes from
plants, animals and microorganisms. Most major
metabolic pathways contain one or more enzyme systems
which are activated by potassium including the
glycolytic pathway, starch synthesis, oxidative

phosphorylation, photophosphorylation, respiration,
nucleotide metabolism and protein synthesis. With most
enzyme systems potassium serves as the most effective
activator, with rubidium or ammonium functioning almost
as effectively. Sodium acts only feebly and lithium
and Tris (hydroxymethyl aminomethane) are
generally completely without effect. While the observed concentration for maximum stimulation varies
with the enzyme studied, a value of around 50 mM
appears to represent an optimal for many systems.
Symptoms of potassium deficiency appear in many plants
when the level of potassium falls below 25 mM, while
the normal concentration of potassium in most plants
is 50 mM or higher. Thus, the concentration of
potassium found in plant tissue corresponds with the
level needed for optimal stimulation of univalent-
cation-requiring-enzyme reactions (Evans and Sorger,
1966).

A few of the key metabolic symptoms of potassium
deficiency in plants include: a) the accumulation of
soluble carbohydrates and reducing sugars; b) amino
acid and amide accumulation; c) blockage of enzymatic
steps associated with starch synthesis, and d) impair-
ment of protein synthesis. Consistent with these
symptoms are the reports that enzyme reactions
associated with protein synthesis require potassium
(Lubin and Ennis, 1963; Schweet, 1964) as well as the
enzyme, starch synthetase, which is involved in starch
metabolism (Akalsuka and Nelson, 1966; Nitsos and
Evans, 1969). Thus, metabolic symptoms of potassium
deficiency can be related to one or more enzyme
systems which have a univalent cation requirement.

Since the capacity of potassium to activate
enzyme systems appears to be a key metabolic function
of potassium in living tissue the question arises as
to the possible mechanism by which potassium or other
univalent cations affect the enzyme reactions. Recent

studies with a number of different univalent cation
requiring enzymes have begun to shed light on at
least some aspects of this question.

## Mechanisms of Potassium Activation With Different Enzyme Systems

Over four dozen enzyme systems have been shown
to be stimulated by univalent cations. Of these
only a handful have been examined from the point of
view of the function of the univalent cation in the
mechanism of action. In no case is the precise
mechanism of action known. Perhaps only when a
univalent-cation-requiring-enzyme has been completely
characterized in terms of amino acid sequence and
3-dimensional structure will it be possible to determine the precise mechanism of univalent cation
activation. Meanwhile the use of a number of different chemical and physical parameters of protein
chemistry have enabled workers to observe differences in the gross structural, chemical and
physical properties of enzymes in a series of different univalent cation environments. These studies
have revealed certain basic features associated with
the function of univalent cations in enzyme activation. The attempt here will be to summarize the
salient information which has been accumulated with
those univalent-cation-enzyme-systems which have
been studied most extensively.

*Pyruvate Kinase*: This enzyme from rabbit muscle
was the first enzyme reported to have a requirement
for univalent cations (Boyer, Lardy, and Phillips;
1942, 1943). It has since been found that the
enzyme from plants and microorganisms share the same
requirement (Seitz, 1949; Miller and Evans, 1957).
Kachmar and Boyer (1953) have shown that the requirement for a univalent cation for pyruvate kinase in

rabbit muscle is satisfied by potassium, rubidium, or
ammonium.  Sodium exhibited a weak capacity to activate, whereas lithium was completely ineffective.
From kinetic considerations they concluded that the
effectiveness of potassium or other univalent cations
to activate the enzyme depends on their capacity to
bind at a negative site(s) and to displace adjoining
structures of the enzyme by some critical amount.
They suggested that the hydrated ionic radii of the
univalent cations may be important in their effectiveness as activators.  However, it has been pointed
out that with the possible exception of lithium, the
hydrated water layer of univalent cations is dis-
placed when a univalent cation is bound to a negative
site on a protein (Becker, 1969).  A more recent
study of univalent cation activation of $\beta$-galacto-
sidase by Becker (1969) has shown no apparent relationship between the degree of activation and the
ionic or hydrated radius (see $\beta$-galactosidase discussion below).

   The technique of immunoelectrophoresis was used
by Sorger, Ford and Evans (1965) to examine the
effect of both activator and non-activator cations on
immunoelectrophoretic behavior of rabbit muscle
pyruvate kinase.  Activator cations such as potassium
or rubidium produced a simple immunoelectrophoretic
pattern containing one major dense reaction band
between antibody and antigen.  In the presence of
non-activator univalent cations such as lithium or
Tris a more complex pattern was obtained which consisted of several additional bands.  From these results it was concluded that the cation environment
affects the structural composition of pyruvate kinase
and thus its activity.  It is not clear precisely
what effect the different univalent cations had on
the protein structure to induce the observed patterns,
but similar experiments with catalase and sheep serum
proteins, both of which do not show a requirement for
a univalent cation, showed no effect on the immuno-

electrophoretic behavior due to the cation environment.

    In a recent attempt to determine the signifi-
cance of the different immunoelectrophoretic patterns,
Betts and Evans (1968a) investigated the effect of
the ionic environments on the electrophoretic
mobility of pyruvate kinase using the moving boundary
method.  The experiments were designed to determine
and compare the mobility of the enzyme in environments
of both KCl and LiCl.  It was found that the enzyme
showed a similar anodic migration in both cationic
environments in the absence of $Mg^{++}$.  In the presence
of $Mg^{++}$ a differential decrease in mobility occurred
so that the enzyme moved less in the non-activator
environment of lithium compared with the activator
environment of potassium.  These results are difficult
to interpret because the presence of $MgCl_2$ may dif-
ferentially influence the binding of potassium and
lithium.  Differential binding would indirectly influence mobility.

    Another technique which has been used to study
the effect of univalent cations on the pyruvate kinase
enzyme is ultraviolet difference spectrophotometry.
Suelter and co-workers reported that a number of
different environmental conditions of crystalline
pyruvate kinase induced changes in the proteins ultraviolet absorption properties (Kayne and Suelter, 1965;
Suelter et al., 1966).  Of interest from the stand-
point of the role of univalent cations in the enzyme
reaction was the comparison of enzyme in TrisCl buffer
containing an univalent cation versus enzyme in
TrisCl buffer and tetramethylammonium chloride.  The
results showed an ultraviolet difference spectrum
characteristic of the perturbation of tryptophanyl
residues.  The perturbation of aromatic residues in
proteins has been interpreted to be due to small
changes in the extinction coefficient and a small
shift in the wavelength of maximum absorption due to

alterations in the chemical and physical environment
of the aromatic ring structure (Yanari and Bovey,
1960).  In the study by Suelter with pyruvate kinase
no attempt was made to correlate the different spectral
results with the known capacity of univalent cations to
function as activators.  More recently, studies by
Wilson, Evans and Becker (1967) examined the effect of
both activator and non-activator ions on the perturbation of tryptophanyl residues.  At physiological
concentrations (0.1M) it was found that only Tris was
able to perturb the residues of tryptophan of pyruvate
kinase.  Of interest in this respect was the observation that the sedimentation co-efficients for
pyruvate kinase in 0.1M KCl was $S^{\circ}_{20w} = 10.08$ but in
0.1M TrisCl was $S^{\circ}_{20w} = 9.55$.  More recently it has
been found that Tris acts as a competitive inhibitor
of potassium binding when present in a reaction medium
with pyruvate kinase (Betts and Evans, 1968b).  If
relatively high concentrations of salt were used in
the difference spectrophotometric studies, other
cations besides Tris were able to induce the ultraviolet difference in absorption.  Thus a comparison
of pyruvate kinase in Tricine buffer and 0.5 M KCl,
an activator, versus enzyme in Tricine buffer and
0.5 M LiCl, a non-activator, gave the characteristic
difference spectra due to perturbation of tryptophanyl
residues (Table I).  At these salt concentrations a
correlation could be made between the capacity of ions
to affect the enzymatic activity and the capacity to
perturb the tryptophanyl residues.  It was concluded
that the structure of the pyruvate kinase molecule was
poised in different conformations by activator and
non-activator salt environments but not in activator
salt environments. As a consequence of the different
initial structures, the addition of higher concentrations
of salts resulted in the perturbation of tryptophan residues in non-activator salt environments but not in activator salt environments.

    Studies have also been conducted to determine the
effect of activator and non-activator univalent

cations on the binding of the substrates, adenosine diphosphate and phospho-enol-pyruvate. The data show no effect on the substrate binding due to the presence of any one of a series of activator and non-activator univalent cations. This indicates that a conformational change induced by univalent cations in pyruvate kinase does not involve an exposer of active sites where substrates bind. It was concluded that the cation requirement is important in some latter stage of the overall mechanism, perhaps in the breakdown of the enzyme-substrate complex (Betts and Evans, 1968a).

It should be pointed out that pyruvate kinase has been shown to exist in a tetramer with a molecular weight of 237,000 a dimer with a molecular weight of 115,000 and a monomer of molecular weight of 57,000 in different urea solutions (Cottam *et al.*, 1969; Steinmetz and Deal, 1966). However, Schilieren patterns of crystalline pyruvate kinase in an analytical ultracentrifuge shows one homogeneous peak of protein regardless of whether the enzyme was dissolved in potassium, lithium or Tris. Thus, no tendency for the enzyme to dissociate into subunits is observed due to the univalent cation environment (Wilson *et al.*, 1967).

*Acetic thiokinase:* This enzymatic reaction proceeds through two steps; one involving the formation of adenyl acetate from acetate and ATP and the second step which involves the formation of acetyl CoA from adenyl acetate and CoA (Berg, 1956; Hiatt, 1964). It has been shown that it is the second of these two steps, the formation of acetyl CoA, which requires an activator cation such as potassium or rubidium and which does not function with non-activator cations sodium or lithium. Kinetic studies by Evans *et al.* (1964) showed that activator cation potassium effects the $V_{max}$ but not the $K_m$ for substrates. Reasoning from this, the authors hypothesized that the most likely reason for a change in the $V_{max}$ was due to a change in the amount of active protein available. This could be achieved if it is assumed that

activator univalent cations induced a conformational
change in a non-active form of the protein to expose
additional active sites.

*Acetaldehyde Dehydrogenase:*  Studies by Sorger
and Evans (1966) show that the activator univalent
cations potassium and rubidium protect acetaldehyde
dehydrogenase from heat inactivation whereas sodium
and lithium were less effective as protectants.
After prolonged dialysis in Tris, a treatment which
leads to enzyme inactivation, the activity of the
enzyme could be renewed by incubation in appropriate
solutions containing univalent cations.  The order
of effectiveness as reactivators was the same as the
order of effectiveness in functioning in enzyme
activity.  The authors suggested that the conformation of the enzyme is more stable in the
presence of the activator than in the non-activator
cations (Sorger and Evans, 1966).

*β-galactosidase:*  Studies with the effect of
univalent cations on this enzyme have utilized its
capacity to function with a number of different substrates, including o-nitrophenyl-β-D-galactopyrano-
side (ONPG), p-nitrophenyl-β-D-galactopyranoside
(PNPG) and lactose to examine the role of univalent
cations in enzyme activation.  Depending on the substrates present, univalent cations had different
effects on enzymatic activity.  Thus, sodium not
only is an ineffective activator of PNPG or lactose
hydrolysis but also competitively inhibits the
potassium activation.  In comparison, studies on the
activation of ONPG hydrolysis by sodium have shown
that the addition of potassium stimulates rather
than inhibits.

In the same studies the effects of hydrostatic
pressures were examined on the catalytic activity of
β-galactosidase in the presence of potassium an activator, and sodium, a non-activator.  From appropriate

calculations it was found that pressures of 1500 atmo-
spheres caused the enzyme volume to increase in the
environment of potassium and decrease in the presence
of sodium (Table 2).  These results were interpreted
to reflect differences due to different conformations
in activator and non-activator univalent cation environments (Becker and Evans, 1969)

     In an attempt to determine the particular
properties of univalent cations which account for differences in activation capacity Becker and Evans (1969)
have examined a series of inorganic univalent cations
including lithium, sodium, potassium, cesium, and
rubidium on the kinetics of β-galactosidase activation.
Based on competitive kinetics it was concluded that
all of these cations appear to bind at the same sites.
It was also determined that no correlation exists between the tenacity of binding of the cations to
β-galactosidase and the capacity to activate.  Thus,
these two processes appear to be separate and distinct
phenomena.  However, it was found that another
property of univalent cations, the effective diameter,
which is defined as the summation of the ionic radius
and hydrated radius of the cation, did correlate with
the activation capacity of different cations.
Potassium, which has the smallest effective diameter
of the ions tested, produced the greatest degree of
activation.  In contrast, lithium, because it may bind
as a partially hydrated form of the ion and thus has
an increased effective diameter, showed no capacity
to activate.  Thus lithium may result in greater
disturbances at the binding site and explain the in-
hibitory effect observed with this enzyme.

     *Formyltetrahydrofolate Synthetase:*  Scott and
Rabinowitz (1967) have shown that this univalent-
cation-requiring-enzyme dissociates into four inactive
subunits in the presence of a non-activator univalent
cation whereas it remains as an catalytically active
monomer in the presence of activator cations.  The

dissociation observed in the absence of an activation
cation could be reversed by the addition of potassium.
The reassociated molecule has almost full enzymic
activity.  More recently it has been reported that the
univalent cation environment has a pronounced effect
on the rate of inactivation in Tris buffer.  Lithium,
sodium, di-, tri- and tetramethylammonium induced more
than 90% loss of enzyme activity in less than 1 hr.,
whereas potassium, rubidium and ammonium retained more
than 50% of the original enzymic activity (Welch *et
al.*, 1968).  It has been suggested that the univalent
cation activation of this enzyme may be due to the
greater accessibility of the substrate binding sites
in the presence of activator univalent cations, perhaps because of the association of the molecule into
one active monomer (Scott and Rabinowitz, 1967).
Consistent with this is the finding that ammonium
increased the binding ($K_m$) of the substrate, formate,
nearly 10 fold (Welch *et al.*, 1968).

*Glycerol dehydrase:*  Schneider *et al.*, (1966)
showed that this univalent cation requiring enzyme
dissociates into two subunits when the enzyme is
passed through a sephadex G-100 column equilibrated
with the non-activator univalent cations, sodium and
cyclohexylammonium.  Neither subunit is enzymatically
active.  Similar experiments in which the sephadex
column has been equilibrated with an activator univalent cation, such as ammonium or potassium results
in one active associated enzyme molecule.  However,
the presence of the glycerol substrate maintains the
enzyme as one monomeric unit even when sodium, a non-
activator univalent cation is present.  Nonetheless,
the enzyme is catalytically inactive which indicates
that the effect of activator univalent cation on this
enzyme is not limited to the subunit of the protein
but must have other functions as well.

*Propionyl-CoA Carboxylase:*  Edwards and Keech
(1968) reported a stimulation in the rate of this

enzymatic reaction due to potassium, but concluded
that there was no absolute requirement for a univalent cation.  They showed that the most significant
kinetic change induced by potassium was a decrease
in the apparent $K_m$ value for the substrate, $HCO_3^-$.
They speculated that the change in the binding constant for $HCO_3^-$ was due to a conformational change in
the enzyme.  Supporting the kinetic data was the reported standard change in entropy of 47.8 entropic
units/per mole due to the presence of potassium.  The
authors concluded that it was difficult to visualise
the nature of the $K^+$ induced conformational change
except that it might substantially influence the
ionic environment at the active site.  This hypothesis was supported by demonstrating that both the
rate of binding of $N-[^{14}C]$ ethylmaleimides, an inacti-
vator of this enzyme, and loss of enzymic activity
caused by N-ethylmaleimide increased in the presence
of potassium.


SIGNIFICANCE OF UNIVALENT CATION ACTIVATING ENZYME
SYSTEMS TO CELLULAR METABOLISM

    It has been previously suggested that a key role
for potassium in cell metabolism may be its capacity
to function in enzyme activation.  The mechanism of
activation by potassium with some enzymes appears to
involve poising the protein molecule in an active
conformation.  With pyruvate kinase this conformation
apparently does not involve the exposer of buried
active sites since substrates can bind with the same
tenacity in the absence of potassium (Betts and
Evans, 1968a).  It appears that with other univalent-
cation-requiring-enzyme systems the cations function
to maintain the enzyme in an aggregated or associated
form (Scott and Rabinowitz, 1967).  Clearly additional studies are needed on other enzyme systems which
require univalent cations to determine the general
nature of these phenomena.

A basic question concerning the general occurrence of univalent-cation-requiring-enzymes is their significance to general cell metabolism. Insight into this question may lie in the observation that potassium can function as an allosteric activator in pyruvate kinase from yeast (Hunsley and Suelter, 1967). In this respect its presence may act to control the metabolic fate of organic acid substrates at this key enzymatic step in glycolysis. Similar control mechanisms could be suggested for other metabolic pathways (Wyatt, 1964).

Important in such a metabolic control mechanism are the factors which regulate the movement and the concentration of univalent cations in cells. In this regard, the extensively studied oubain-sensitive-ATPase which is activated by both potassium and sodium undoubtedly plays an important role in univalent cation movement at cell membranes (Lowe, 1968). Also it has been shown that intimately associated with energy conserving processes of cell organelles including mitochondria (Lehninger, 1967) and chloroplasts (Packer and Crofts, 1967) there are ion transport systems which function in the movement and compartmentization of univalent cations. Further research studies may show that a fundamental mechanism for regulation of metabolic events involves a highly integrated control system of enzyme activator univalent cations and other metabolic constituents in living cells. Such a system would undoubtedly play a major role in the growth, development and well being of living organisms.

TABLE 1.  The Effect of Different Univalent Cations at
          a Series of Concentrations on the Absorbancy
          Differences at 295 mu of Pyruvate Kinase
          (after Wilson *et al*., 1967)*

| Ion Comparison | | Absorbance Difference at 295 mu at various concentrations of cation chloride ($-O.D.X\ 10^3$) | |
|---|---|---|---|
| Sample Cuvette | Reference Cuvette | | |
| | | 0.5M | 0.86M |
| K+ | Li+ | 5 | 17 |
| Rb+ | Li+ | 7 | 18 |
| K+ | TMA+ | 4 | 16 |
| K+ | Na+ | 2 | 7 |
| Na | Li+ | 3 | 13 |
| K+ | Rb+ | 0 | 0 |

*Each 3 ml reaction contained 0.33 M Tricine buffer
pH 7.4, $6.7 \times 10^4$ or mercaptoethanol, 6 mg of
pyruvate kinase and salts as indicated.

TABLE 2.   The Effect of Hydrostatic Pressure on the
           Enzymatic Hydrolysis of Lactose by β-ga-
           lactosidase (After Becker and Evans, 1969).

| Cation Addition | Cation Conc. (mM) | $\Delta V$ (cm$^3$/mole) |
|:---:|:---:|:---:|
| K+ | 45 | +4.5 |
| Na+ | 20 | -12.0 |

The incubation mixture (4 ml) consisted of 0.1 M
histidine, 45 mM lactose, 5 mM dithiothreitol, the
enzyme preparation (72 ug protein) and the
appropriate cation chloride addition.   The assay
mixtures were incubated at 5° for 1 hr.   (Repro-
duced with permission from Elsevier Publishing
Company.)

REFERENCES


1.    AKATSUKA, T. and NELSON, O.E. (1966)  Granule-
      bound adenosine diphosphate glucose-starch
      glucosyl transferase of maize seeds.  *J. Biol.
      Chem. 241*: 2280-86.

2.    BECKER, V.E.  Monovalent cations and β-galactosi-
      dase.  *Arch. of Biochem. and Biophys.*  (In
      press.)

3.    BECKER, V.E. and EVANS, H.J. (1969)  The influence
      on monovalent cations and hydrostatic pressure
      on β-galactosidase activity.  *Biochem. Biophys.
      Acta. 191*: 95-104.

4.    BERG, P. (1956)  Acyl adenylates: an enzymatic
      mechanism of acetate activation.  *J. Biol. Chem.
      222*: 991-1013.

5.    BETTS, G.F. and EVANS, H.J. (1968a)  The effect of
      cations on the electrophoretic mobility and sub-
      strate binding properties of pyruvate kinase.
      *Biochem. Biophys. Acta. 167*: 190-193.

6.    ——————————————————— (1968b)  The inhibition
      of univalent cation activated enzymes by tris
      (hydroxymethyl aminoethane).  *Biochem. Biophys.
      Acta. 167*: 193-196.

7.    BOYER, P.D., LARDY, H.A. and PHILLIPS, P.H. (1942)
      The role of potassium in muscle phosphorylations.
      *J. Biol. Chem. 146*: 673-682.

8.    ——————————————————————————————— (1943)
      Further studies on the role of potassium and
      other ions in the phosphorylation of the adenylic
      system.  *J. Biol. Chem. 149*: 529-541.

9. COTTAM, G.L., HOLLENBERG, P.F. and COON, M.J. (1969) Subunit structure of rabbit muscle pyruvate kinase. *J. Biol. Chem. 241*: 1481-1486.

10. EDWARDS, J.B., KEECH, D.B. (1968) Activation of pig heart propionyl-CoA carboxylase by potassium ions. *Biochem. Biophys. Acta. 159*: 167-175.

11. EPSTEIN, E. (1965) Mineral Metabolism. In: *Plant Biochemistry*. Bonner, J. and Varner, J.E. (eds.). pp. 438-66. Academic Press, N.Y.

12. EVANS, H.J. and SORGER, G. (1966) Role of mineral elements with emphasis on the univalent cations. *Ann. Rev. Plant Physiol. 17*: 47-76.

13. EVANS, H.J., CLARK, R.B. and RUSSELL, S.A. (1964) Cation requirements for the acetic thiokinase from yeast. *Biochem. Biophys. Acta. 92*: 582-594.

14. HIATT, A.J. (1964) Further studies on the activation of acetic thiokinase by magnesium and univalent cations. *Plant Physiol. 39*: 475-479.

15. HUNSLEY, J.R. and SUELTER, C.H. (1967) Allosteric properties and preparation of homogeneous yeast pyruvate kinase. *Fed. Proc. 26*: 559.

16. KACHMAR, J.F. and BOYER, P.D. (1953) Kinetic analysis of enzyme reactions. II: The potassium activation and calcium inhibition of pyruvic phosphoferase. *J. Biol. Chem. 200*: 669-82.

17. KAYNE, F.J. and SUELTER, C.H. (1965) Effects of temperature, substrate and activating cations on the conformation of pyruvate kinase in aqueous solution. *J. Amer. Chem. Soc. 87*: 897-901.

18.  LEHNINGER, A.L., CARAFOLI, E. and ROSSI, C.S. (1967)
     Energy-linked ion movements in mitochondrial
     systems. *Ad. Enzymol. 29*: 259-320.

19.  LOWE, A.G. (1968)  Enzyme mechanism for the active
     transport of sodium and potassium ions in ani-
     mal cells. *Nature 219*: 934-36.

20.  LUBIN, M. and ENNIS, J.H. (1963)  The role on in-
     tracellular potassium in protein synthesis.
     *Federation Proc. 22*: 302.

21.  MILLER, G. and EVANS, H.J. (1957)  The influence
     of salts on pyruvate kinase from tissues of
     higher plants. *Plant Physiol. 32*: 346-54.

22.  NASON, A. and McELROY, W.D. (1963)  Modes of ac-
     tion essential mineral elements.  In: Steward,
     F.C. (ed.), *Plant Physiol. III*: 449-536.

23.  NITSOS, R.E. and EVANS, H.J. (1969)  Effects of
     univalent cations on the activity of particu-
     late starch synthetase. *Plant Physiol. 44*:
     1260-1266.

24.  PACKER, L. and CROFTS, A.R. (1966)  The energized
     movement of ions and water by chloroplasts.
     *Current Topics in Bioenergetics  2*: 24-64.

25.  SCHNEIDER, S., PECK, K. and PAWELKIEWICZ, J. (1966)
     Enzymic transformation of glycerol to β-hydro-
     xypropionic aldehyde.  II: Dissociation of the
     enzyme into two protein fragments. *Bull. Acad.
     Polonaise Sci. XIV*: 7-12.

26.  SCHWEET, R.S., ARLINGHAUS, R., SHAEFFER, J. and
     WILLIAMSON, A. (1964)  Studies on hemoglobin
     synthesis. *Medicine 43*: 731-45.

27.  SCOTT, J.M. and RABINOWITZ, J.C. (1967)  The as-
     sociation-dissociation of formytetrahydrofolate
     synthetase and its relation to monovalent cation
     activation of catalytic activity.  *Biochem.*
     *Biophys. Res. Commun. 29*: 418-23.

28.  SEITZ, I.F. (1949)  Role of potassium and ammonium
     in the transfer of phosphate from phosphopyruvic
     acid to the adenylic system.  *Biokhimiya 14*:
     134-140.

29.  SORGER, G.J., FORD, R.E. and EVANS, H.J. (1965)
     Effect of univalent cations on the immunoelectro-
     phoretic behavior of pyruvate kinase.  *Proc.*
     *Nat. Acad. Sci. U.S. 54*: 1614-1621.

30.  SORGER, G.J. and EVANS, H.J. (1966)   Effect of
     univalent cations on the properties of yeast
     NAD+ acetaldehyde dehydrogenase.  *Biochem. Bio-*
     *phys. Acta. 118*: 1-8.

31.  STEINMETZ, M.A. and DEAL, W.C., Jr. (1966) Meta-
     bolic control and structure of glyolytic en-
     zymes.  II: Dissociation and subunit structure
     of rabbit muscule pyruvate kinase.  *Biochem. 5*:
     1399-1405.

32.  SUELTER, C.H., SINGLETON, R. Jr., KAYNE, F.J.,
     ARRINGTON, S., GLASS, J. and MILDVAN, A.S. (1966)
     Studies on the interaction of substrate and
     monovalent and divalent cations with pyruvate
     kinase.  *Biochemistry 5*: 131-139.

33.  SUTCLIFFE, J.F. (1962) *Mineral Salts Absorption in*
     *Plants*. N.Y., Pergamon Press, p. 194.  (Inter-
     national Series of Monographs on Pure and Ap-
     plied Biology.  Division: Plant Physiology Vol.
     1.)

34.  WELCH, W.H., IRWIN, C.L. and HIMES, R.H. (1968)
     Observations on the monovalent cation require-
     ments of formyltetrahydrofolate synthetase.
     *Biochem. Biophys. Research Comm. 30*: 255-261.

35.  WILSON, R.H., EVANS, H.J. and BECKER, R.R. (1967)
     The effect of univalent cations salts on the
     stability and on certain physical properties
     of pyruvate kinase. *J. Biol. Chem. 242*: 3825-
     3832.

36.  WYATT, H.V. (1964)  Cations, enzymes and control of
     cell metabolism. *J. Theoret. Biol. 6*: 441-470.

37.  YANARI, S. and BOVEY, F.A. (1960)  Interpretation
     of the ultraviolet spectral changes of proteins.
     *J. Biol. Chem. 235*: 2818-2826.

*Questions to Dr. Wilson*

BAR-AKIVA:  You have shown that magnesium was one of
the cations which affect enzymes. But in your further
experiments you seem to have used only univalent ions.
Can you give us any additional information? We have
done some experiments in this direction and it seems
to us that magnesium and possibly, also manganese, are
very highly effective in activating this enzyme system.

WILSON:  Korn and Kelkar of Pennsylvania have shown
that magnesium is functioning and complexing with the
ADP molecule. We have in fact done some studies with
divalent ions, but I didn't have time to go into them
today.

STEWARD:  At what concentration levels do you use the
activating cations; do they bear any rational relationship with the concentration levels that might con-
ceivably be obtained in the cell?

WILSON:  Yes, I pointed out in suggesting that this may
be a major function of potassium, the level of which in
a cell probably runs around .05 molar. Certainly there
is a considerable variability in the reported literature
in terms of potassium concentrations, but as a rough
estimate we can take that - and in most enzyme systems
which require potassium for activity - this is a type of
concentration at which you're reaching your plateau level
in terms of catalytic function. So it looks like there's
a fairly good correlation there.

STEWARD:  Do you have any idea of what the concentration
is in the cytoplasm?

WILSON:  It's hard to say exactly what the concentration
is in the cytoplasm, but we have to base it primarily on
the studies which have been done reporting the potassium
concentration in a plant in general. In fact, there's
good evidence now coming out of the animal mitochondrial
studies and in chloroplasts as well, that potassium can
be compartmentalized in different organelles. This may
have an influence on the regulatory mechanisms of potassium activity. It may be that a compartmentalization of
potassium plays a role in regulating, when enzymes are
going to be turned on and off by univalent cations.

HOROWITZ:  Estque vous avez etudié aussi le sodium?

WILSON:  In most cases sodium acts as a very weak
activator of enzymes of normal plants. In the case of
some halophites this is reversed, in that sodium is
functioning as a primary activator.

# FREE AMINE CONTENT IN FRUIT TREE ORGANS
# AS AN INDICATOR OF THE NUTRITIONAL STATUS
# WITH RESPECT TO POTASSIUM *

## Moshe Hoffman and Rudolph M. Samish

*Division of Pomology, Viticulture and Oleoculture,
Volcani Institute of Agricultural Research, Bet Dagan,
Israel*

## ABSTRACT

A survey of free nitrogenous compounds other than
amines (F.N.C.) in leaves of apple, apricot, peach and
grape vine, grown at various levels of potassium (from
deficiency to excess which retarded growth) and with
varying calcium supply.  Optimum nutrient supply was
associated with low levels of F.N.C., while both deficiency and excess of potassium supply, caused an
increase in F.N.C. content in the leaves of all species
under study.  Excess of calcium did not influence the
F.N.C. content of leaves and when high calcium accompanied excess potassium supply it annuled the effect
of high potassium on F.N.C. in the leaves.

Poisonous amines (putrescine and agmatine) were
found to be associated very closely with potassium.

* Contribution from The Volcani Institute of Agricultural Research, Bet Dagan, Israel.  1970 Series, No.
1735 -E.

Moshe HOFFMAN. Sen. Pomol., Volcani Inst. Agric. Res., Bet Dagon (Israel) since 1965.  b. 1924 Haifa
    (Israel); 1956 B.Sc., Univ. Calif.; 1958 M.Sc., State Univ. Washington; 1970 Ph.D., Heb. Univ.
    Jerusalem-Rehovot; 1958 Res. Pomol., Volcani Inst. Agric. Res., Rehovot (Israel).

nutrition of fruit trees.  Their content increased both
under conditions of potassium deficiency and excess, and
their level was not influenced by calcium nutrition.
These amines were observed both during the growing period
in the leaves and during dormancy in the bark and in
root tips of fruit trees and vines, and thus provide a
means to detect potassium supply level in fruit trees
throughout the year.  These amines were not detected
under conditions of deficiency or excess of elements
other than potassium.

A very simple and rapid method for the detection
of agmatine and putrescine in plant material was developed, which makes it possible to evaluate the nutritional status of potassium in fruit trees.

INTRODUCTION

The accepted method of evaluating the potassiun nutritional status of fruit trees is by their response to
the application of various amounts and forms of fertilizers, and by assessing the potassium content in their
leaves.  The first method is time consuming and costly,
and the results are relevant for only one particular lo-
cality.  The second method determines the availability,
rather than the sufficiency or excess of the element.
The presence of potassium in the leaf is influenced by
many internal and external factors, which render evaluation of its nutritional status even more difficult
than with a number of other elements (Smith, 1965).

While searching for more convenient and precise
methods for evaluating the potassium nutrition status
of plants, it was observed that both deficiency and excess of potassium supply induce the accumulation of free
amino compounds in plant leaves (Richards and Templeman,
1936).  However, the pattern of accumulation of these
compounds is altered by external conditions (Lewis

et al., 1964; Steward et al., 1959).

Richards and Coleman in 1952 (Richards and Coleman, 1952) showed that free poisonous amines (putrescine and agmatine) accumulate in potassium-deficient barley leaves. Subsequently, other workers found these compounds in the leaves of several other plant species grown under conditions of potassium deficiency (Freiberg and Steward, 1960; Smith, 1967), but not in every case (Freiberg and Steward, 1960).

At the previous meeting of this forum in 1966, we reported on the presence of these amines in potassium-deficient fruit tree and grapevine leaves. It was shown that the content of amines was affected by season and by leaf age, being especially high in young leaves at the beginning of the growing season, and dropping in mature leaves and also later in the season (Samish and Hoffman, 1966).

The present report describes the usefulness of the determination of putrescine in fruit tree and grapevine leaves, for evaluating the potassium nutrition status of such plants.

MATERIALS AND METHODS

*Plant Material*

Apricot and apple seedlings and short grapevine cuttings were grown in 10-liter plastic pails in a semi-automatic sand culture system [a modification of the set-up described by Eaton and Bernardin (Eaton and Bernardin, 1962)].

The plants were grown in a nutrient solution based on Hoagland and Arnon's recommendation (Hoagland and

Arnon, 1938); the solutions differed in their levels of
potassium and calcium.  On the basis of the findings, we
determined the deficient, optimal and supraoptimal K
concentrations in the nutrient solution, at different
Ca levels, for each of the tested species.

*Analytical Methods*


    Leaf samples for inorganic mineral analyses were
dried at 60°C in a forced-air oven.  Potassium was determined after wet digestion (Cotton, 1962) with a Uni-
cum Atomic Absorption SP 90 spectrophotometer.

    For analyses of organic compounds, leaves were ex-
tracted fresh, immediately after sampling.  Amino compounds were extracted as suggested by Steward (Steward
et al., 1959) and Plaisted (Plaisted, 1958) and separated
by two-dimensional chromatography (Kliewer and Nasser,
1966).  The ninhidrin-stained spots were eluted and
evaluated quantitatively with a Spectronic 20 spectrophotometer.

    For amine extraction, a modification of the method
suggested by Smith (Smith, 1963) was used.  Finely cut
plant leaves were immersed overnight in 0.01N $H_2SO_4$
(1:10, W:V).  The leaf extract was applied directly on
to 30 × 250 mm Whatman 3-mm strips at an amount equi-
valent to 10-20 mg plant material.  Ascending chromatography was used with a mixture of methyl ethyl-ketone:
n-butanol: $NH_4OH$ (26%): water (5:3:1:1).

    With this solvent system, the Rf value for agma-
tine is about 0.3 and that for putrescine about 0.6,
but its spot is rather diffused.  Quantitative determinations of the two amines were carried out either by
scanning (with a Photovolt densitometer), or by comparing the spots with those of chromatogramed standards
of the amines.

RESULTS

The relative heights of the three species grown in
different nutrient solutions is presented in Fig. 1.
The data produced optimum curves, the peaks of which appeared at different potassium concentrations for different species.  These optima shifted to higher K values
when the Ca level in the nutrient solution was raised.
A potassium supply above the optimum resulted in reduced
plant growth.

The potassium content in the leaves was proportional to potassium supply even beyond the optimum,
but the Ca level in the nutrient solution had very little
effect on potassium accumulation (Fig. 2).

Total amino acid and amide contents of the leaves
are shown in Table 1.  These compounds tend to accumulate in leaves under conditions of K deficiency or excess.  Calcium excess does not affect the accumulation
of free amino compounds and, when occurring along with
potassium excess, annuls the effect of high K supply (at
the concentration used) on the accumulation of these compounds.  This general trend of the accumulation of
amino compounds was found to be the same in the leaves
of plants grown in the open and in a greenhouse, but
different compounds showed different patterns of accumulation under these two growth conditions.  The effect of
the nutrient treatments on the accumulation of amides
(asparagine and glutamine) was more pronounced.  These
compounds tend to accumulate under conditions of restricted growth in general, and the trend does not seem
to be connected with K nutrition in particular.

Fig. 3 shows paper chromatograms from leaf extracts
of plants grown under +K and -K conditions.  The appearance of poisonous amines is very striking and reveals
their large content in the potassium-deficient plants.

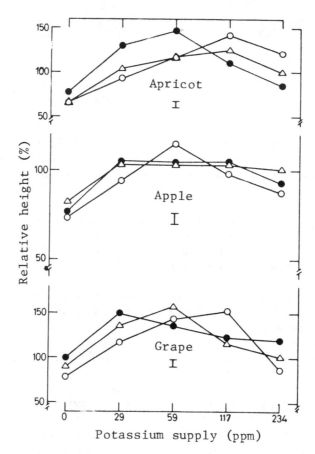

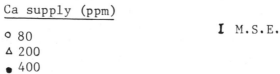

FIG. 1.  Relative height of plants grown in nutrient
         solutions containing varying amounts of
         potassium and calcium

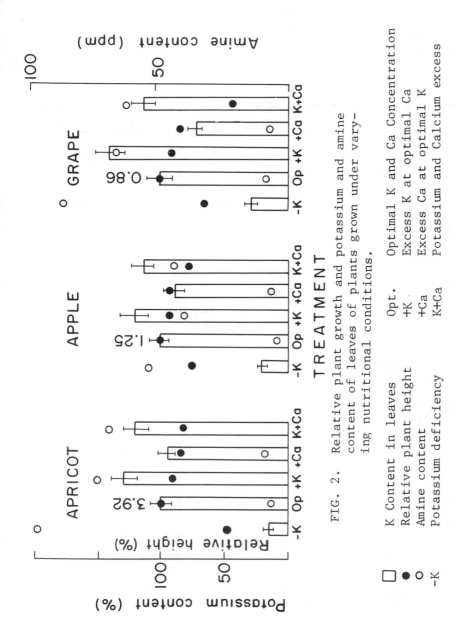

FIG. 2.  Relative plant growth and potassium and amine
         content of leaves of plants grown under vary-
         ing nutritional conditions.

| K Content in leaves | Opt. | Optimal K and Ca Concentration |
| Relative plant height | +K | Excess K at optimal Ca |
| Amine content | +Ca | Excess Ca at optimal K |
| Potassium deficiency | K+Ca | Potassium and Calcium excess |

□  ●  ○  -K

TABLE 1

Amount of Free Amino Compounds[1] in Leaves of Plants
Grown under Various Nutritional Conditions (as ppm
Leucine)

| Treatment K: | deficient | optimum | excess | optimum | excess |
|---|---|---|---|---|---|
| Ca: | | normal | | excess | |
| *Apple* | | | | | |
| Total free amino compounds | $406^{b}$ [2] | $322^{a}$ | $381^{b}$ | $280^{a}$ | $356^{ab}$ |
| Asparagine + Glutamine | $67^{c}$ | $-^{a}$ | $14^{b}$ | $-^{a}$ | $10^{b}$ |
| *Apricot* | | | | | |
| Total amino compounds | $2268^{b}$ | $1423^{a}$ | $2406^{b}$ | $1412^{a}$ | $1572^{a}$ |
| Asparagine + Glutamine | $637^{b}$ | $273^{a}$ | $1000^{c}$ | $503^{b}$ | $520^{b}$ |
| *Grapevine* | | | | | |
| Total amino compounds | $3278^{c}$ | $1458^{a}$ | $2685^{bc}$ | $1465^{a}$ | $2139^{b}$ |
| Asparagine + Glutamine | $751^{b}$ | $563^{a}$ | $762^{b}$ | $572^{a}$ | $720^{b}$ |

1)  Not including amines.
2)  No significant difference between numbers carrying
    common letter within the same species.

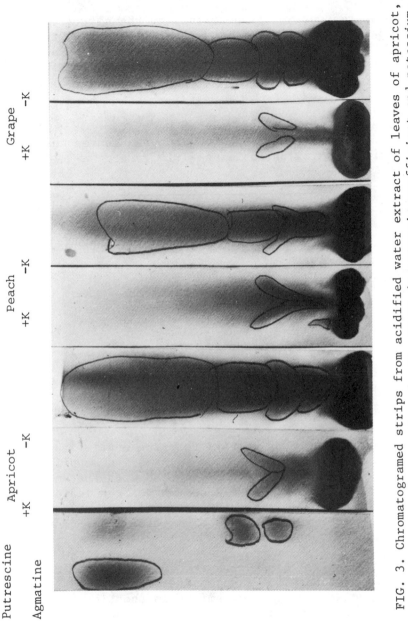

FIG. 3. Chromatogramed strips from acidified water extract of leaves of apricot, peach and grapevine plants grown in potassium-sufficient and potassium-deficient nutrient solution.

The amine content was proportional to the severity of
the K-deficiency of the plants.  Semillion grapevine
leaves, collected from rootstock experiments, were
tested for their K and amine content (Table 2).

TABLE 2

Potassium and Amine Content in Grapevine Leaves Collected from Semillion Rootstock Trial.

| K%(%) | Amines (ppm) |
|---|---|
| 0.65 a[1] | 50 a |
| 0.53 a | 50 a |
| 0.42 ab | 120 b |
| 0.35 b | 520 c |
| 0.28 b | 950 d |

[1] No significant difference between values
carrying a common letter.

The data indicate that in this case the statistical dis-
tinction between deficiency (as well as its severity)
and critical level was more pronounced with the amine
determination than with the K content of the leaves.
Therefore, the amine content served as a better tool for
the determination of the potassium nutritional status
of the vines.

     In a parallel study with *in vitro* apple callus tis-
sues, it was found that excess potassium supply also
induces the accumulation of amines in the tissues.
Checking the apple leaves grown in solutions with

different potassium levels, it was found that amines accumulated also at excess levels of potassium supply
(Fig. 4), i.e., the amine content in apple leaves was
high not only in K-deficient but also in K-excess
plants.  The amine content of the leaves was hence inversely related to the growth rate of the plants and
served, therefore, as a good indicator for the growth
response to be expected from differential potassium
nutrition.


DISCUSSION


In evaluating potassium nutrition of plants, results from numerous experiments were always correlated
with plant performance and the potassium content of
plant leaves (Forshey, 1969).  The present results also
show that with the aid of these two parameters we can
establish the critical K content in the leaf and demon-
strate their dependence upon other factors (such as
genetic constitution and Ca supply).  We confirmed other
work (Hewitt, 1963) showing that the quantity of free
amino compounds in fruit tree leaves can point to nutritional disorders in the plant.

The amine content of leaves was found to be more
specific, less cumbersome and more useful in evaluating
K nutrition of fruit trees, because it is possible, by
manipulating the K supply, to aim at the desired situation where amine content will be kept at a minimum
level.

To determine the specificity of the test, we
checked the leaves from apple trees suffering from Zn,
Fe, Mg, and Mn deficiencies, and grapevine and pecan
leaves suffering from $Na^+$ and $Cl^-$ salinity, but could
not trace poisonous amines in them.  However, recent
reports (Sinclair, 1967) show that some agmatine is

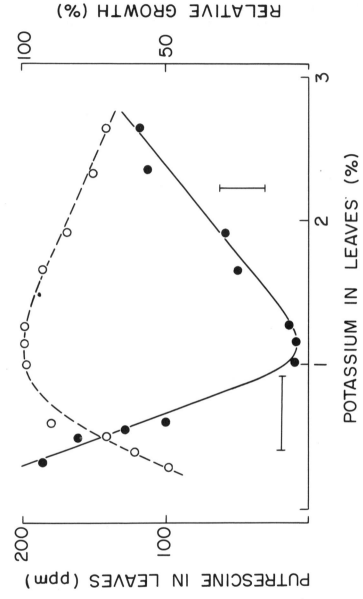

FIG. 4.   Potassium and putrescine concentration in apple
leaves and relative growth of apple seedling
grown in nutrient solution.

accumulated in barley leaves severely deficient in Mg
and P.  Main attention should therefore be paid to putrescine, rather than to agmatine formation, when in-
vestigating K nutrition.

As for the significance of amine formation under
conditions of unbalanced K nutrition, it seems that
free amines are produced in plants that contain an excess of free organic acids (Smith, 1967), neutralizing
the acids and thus maintaining a constant pH value of
the cell sap.  A few nutritional disorders induce the
accumulation of organic acids in the cell (David, 1959),
but potassium malnutrition induces this effect more
and faster than other elements (Hewitt, 1963).  It is
also possible that the primary effect of potassium malnutrition is upon the rapid accumulation of citric acid
[A. Gur (unpublished data) found acetic acid in the
roots] which induces the amine formation (Coil, 1948;
Cummings and Teel, 1965), while for example lime-
induced iron deficiency causes the accumulation of
malic acid (Wallace and Lunt, 1960), which does not trig-
ger this mechanism.

We have shown that the accumulation of amines in
deciduous fruit plants is quantitatively related to
both sub- and supra-optimal K nutrition; hence, amine
analysis can be used to determine the K nutritional
status of such plants.

The hitherto almost exclusively used method of
foliar inorganic analysis of the K concentration, which
provides absolute data, is less useful, inasmuch as
physiologically significant levels have to be determined beforehand in order to interpret the analysis of
salts in terms of performance.  We have shown again
that the physiologically optimal K concentration will
differ according to the genetic constitution and environmental conditions of the plant.  The relative
amine concentration is in itself a physiological response and therefore directly indicative of the

nutritional status, which affects plant performance.
The method is relatively easy to employ, with simple
equipment, and has also the considerable advantage that
it can (with deciduous trees) be carried out even in
winter, thus enabling fertilizer application for the
coming season.

## REFERENCES

1.    COIL, B.J. (1948) Potassium deficiency and excess
      in Guyaule. *II*: Cation-anion balance in the
      leaves. *Pl. Physiol. 23*: 403-424.

2.    COTTON, R.H. (1962) Determination of nitrogen,
      phosphorus and potassium iń leaf tissue. Ap-
      plication of micromethods. *Ind. Eng. Chem.
      Analyt. Edn. 17*: 734-738.

3.    CUMMINGS, G.A., and TEEL, M.R. (1965) The effect
      of nitrogen, potassium and plant age on cer-
      tain nitrogenous constituents and malate con-
      tent of orchard grass (Dactylis glomeratah).
      *Agron. J. 57*: 127-129.

4.    DAVID, D.D. (1959) Organic acid metabolism in
      plants. *Biol. Rev. 34*: 404-444.

5.    EATON, F.M. and BERNARDIN, J.E. (1962) Soxhlet
      type automatic sand culture. *Pl. Physiol.
      37*: 357.

6.    FORSHEY, C.J. (1969) Potassium nutrition of de-
      ciduous fruits. *Proc. of Potassium in Horti-
      culture, Symp. Am. Soc. Hort. Sci. 1968*,
      Davis, Calif., pp. 7-9.

7.    FREIBERG, S.R. and STEWARD, F.G. (1960) Physio-
      logical investigations on the banana plant.
      *III*: Factors which affect the nitrogen com-

pounds of the leaves. *Ann. Botany. (London) 24*: 247–257.

8.  HEWITT, E.J. (1963) The essential nutrient elements. Requirements and interactions in plants. In: *Plant Physiology III*: 137–360, (ed.) F.C. Steward. Academic Press, N.Y.

9.  HOAGLAND, D.R. and ARNON, D.I. (1938) The water culture method of growing plants without soil. *Circ. Calif. Agric. Exp. Stn. 347*.

10. KLIEWER, W.M. and NASSER, A.R. (1966) Changes of concentration of organic acids, sugars and amino acids in grape leaves. *Am. J. Enol. Viticul. 17*: 48–57.

11. LEWIS, L.N., TOLBERT, N.E. and KENWORTHY, A.L. (1964) Influence of mineral nutrition on the amino acid composition of Bartlett pear tree. *Proc. Am. Soc. Hort. Sci. 83*: 185–194.

12. PLAISTED, P.H. (1958) Clearing free acid solutions of plant extract for paper chromatography. *Contrib. Boyce Thompson Inst. Pl. Res. 19*: 231–244,

13. RICHARDS, F.J. and COLEMAN, R.G. (1952) Occurrence of putrescine in potassium-deficient barley. *Nature 170*: 460.

14. RICHARDS, F.J. and TEMPLEMAN, W.G. (1936) Physiological studies in plant nutrition. IV: Nitrogen metabolism in relation to nutrient deficiencies and age in leaves of barley plants. *Ann. Botany (London) 50*: 307–402.

15. SAMISH, R.M. and HOFFMAN, M. (1966) Free nitrogenous compounds as an indicator for potassium nutrition of fruit trees. Avstract in: *Proc. 17th Int. Hort. Cong.*, Baltimore, Md.

16.  SINCLAIR, C. (1967) Relation between mineral de-
     ficiency and amine synthesis in barley. *Nature*
     *213*: 214-215.

17.  SMITH, P.F. (1965) Mineral analysis of plant tis-
     sue. *Ann. Rev. Pl. Physiol. 13*: 81-107.

18.  SMITH, T.A. (1963) L-arginine carboxy-layse of
     higher plants and its relation to potassium
     nutrition. *Phytochemistry 2*: 241-252.

19.  _____ (1967) The biosynthesis of putrescine
     in higher plants and its relation to potassium
     nutrition. *Potash Rev. Sub. 3/25*: 1-8.

20.  STEWARD, F.C., CRANT, F., MILLAR, K., ZACHARIS,
     R.M., ROBSON, B. and MARGOLIS, P. (1959) Nu-
     tritional and environmental effects on the
     nitrogen metabolism of plants. In: *Utilization
     of Nitrogen and its Compounds by Plants. Symp.
     Soc. Exp. Biol. 13*: 148-176.

21.  WALLACE, A. and LUNT, O.R. (1960) Iron chlorosis
     in horticultural plants (a review). *Proc. Am.
     Soc. Hort. Sci. 75*: 819-841.

*Question to Mr. Hoffman*

POSSINGHAM:  Do putrescine and agmatine accumulate
under other restraints than that of potassium, that
affect growth, like low temperature, or moisture
stress?

HOFFMAN:  High temperature apparently does affect the
accumulation of putrescine, because we found its level
pretty high in the early season in the leaves, but
later on, in summer when both air and root temperatures rise, free amine levels go down.

As to moisture stress there exists a report by Strobonov from Russia that if a high concentration of sodium chloride is put on the leaf of cotton plants, it induces putrescine formation. The species that I dealt with didn't show the same effect. I took plant material from salt tolerance experiments whose leaves were tremendously high in sodium chloride and sulfate but they didn't accumulate putrescine.

I couldn't be certain as to the specificity of agmatine, because late reports show that you can get some traces of agmatine under a very severe phosphorus and magnesium deficiency. What is apparently specific for potassium nutrition is the later deterioration of agmatine into putrescine through the metabolic pathway, which is blocked upon satisfactory level of potassium supply.

STEWARD:  Are there any other amines that behave similarly, the volatile ones for example?

HOFFMAN:  I didn't check, but I know that somebody at Cornell University did trace some cadaverine in potassium deficient leaves of apple. There are two metabolic pathways for putrescine formation in nature. The substrate apparently is arginine, which can go into citralline and later to ornithine and finally to putrescine. The other one is from arginine to agmatine, further through N carbamyl-putrescine to putrescine.

The first metabolic pathway consists of deamination and later decarboxylation; the second one decarboxylation and later on deamination. The pathway was studied by Smith and co-workers at the Wye Institute and they found that the decarboxylation-deamination pathway is the one existing in barley. We found that the formation of putrescine in fruit trees is also through decarboxylation and subsequent deamination.

In some plants, apparently, putrescine doesn't
accumulate, and deteriorates further to volatile compounds, presumably $CO_2$ and ethylene. If you take, for
example, bean or pea plants you wouldn't find the amines,
they degrade further. But in our fortunate case they do
accumulate.

The site of formation of these amines is primarily
in the young tissue near the growing points, but not
only there. We can detect them in the bark and in the
roots; sometimes in the root even at higher concentration than in the leaf. As a matter of fact, this method
makes one independent of the season. You can ·check
trees for their potassium nutrition, by analyzing the
bark in mid-winter. Or you can check them very early in
the season when the amount of potassium in young leaves
is still a little bit high, and way before any K deficiency symptoms will appear on the tree.

The fact that putrescine accumulates both under low
and high K nutrition but disappears under optimum conditions, is a rather good answer to the quest for an indicator that will show us the best growing condition of
the plant. This test does hold true in the field. I
checked it all over the country, and it's very easy to
perform.

GUR:  We made essays by gas chromatography on K deficient
apple leaves and would have detected other volatile
amines, but we did find only this putrescine.

As to the specificity of putrescine, we shall show
evidences indicating that putrescine may probably occur
under the influence of certain acids, and Smith has shown
that acid feeding may cause the production of putrescine.
Now it may be asked under what conditions do these acids
appear? So far we know only that potassium deficiency
causes the appearance of certain acids.

# K+ RECIRCULATION IN PLANTS AND ITS IMPORTANCE FOR ADEQUATE NITRATE NUTRITION

S. Herman Lips, A. Ben-Zioni

*Dept. of Plant Physiology, Negev Institute for Arid Zone Research, Beersheva, Israel*

and Yoash Vaadia

*Volcani Institute for Agricultural Research, Bet Dagan, Israel*

## ABSTRACT

$KNO_3$ is taken up by the roots of plants and trans-located to the shoot where it is reduced and incorpor-ated into organic compounds. The reduction of $NO_3^-$ in the shoot is accompanied by a concomitant synthesis and accumulation of malic acid. The amount of malate found is less than half the amount of reduced nitrogen in the shoot in long term experiments. It has been suggested that part of the malate is translocated as the potassium salt from the shoot to the root, malate (or its degradation product bicarbonate) would exchange with nitrate. Under these circumstances $K^+$ would move upwards as $KNO_3$ and downwards as $K^+$ malate. Several types of experiments support this hypothesis: (a) $NO_3$ uptake by tobacco plants decreases with decreasing nitrate reduction - less malate is available for ex-change; (b) net $K^+$ uptake increases with decreasing

S. Herman LIPS. Sen. Plant Physiol., Negev Inst. Arid Zone Res., Beersheba (Israel). b. 1932 La Plata (Argentina); 1960 B.Sc. (Agr.), Heb. Univ., Rehovot (Israel); 1961 M.Sc. and 1964 Ph.D., Univ. Calif., Los Angeles (U.S.A.).

nitrate reduction - a smaller pool of endogenous $K^+$ participates in exchange reactions with external $K^+$; (c) the ratio of $NO_3^-/K^+$ in net uptake decreases with decreasing nitrate reduction while the same ratio in the exudate remains constantly close to one. The metabolic requirements of nitrogen control to a large extent the rate of nitrate uptake. This control is achieved by the production of malate in the shoot and its translocation to the root where it exchanges for nitrate.

---

$K^+$ may move in plants from root to shoot as $KNO_3$ and from shoot to root as $K^+$-malate. Malate is formed in leaves in response to the reduction of nitrate. After its translocation to the root, malate is oxidized and $HCO_3^-$ is given off to the medium being exchanged for $NO_3^-$ uptake by roots by the rate of nitrate reduction in the leaves.

MATERIALS AND METHODS

Nitrate was determined by the method of Sloan and Sublett.

$K^+$ was determined on a Unicam Atomic Absorption apparatus.

Tobacco (*Nicotiana rustica* L.) was germinated and grown in half-Hoagland solution.

Nitrate reductase activity was assayed according to Bar Akiva and Sagiv. Crude extracts were used. Mercaptoethanol replaced cystein in the extraction medium.

RESULTS AND DISCUSSION

The synthesis of malate accompanying nitrate reduction has been shown by Ben Zioni *et al.* (1970).
Leaves from plants kept in solutions containing nitrate
divert more label from $^{14}CO_2$ to malate during photosynthesis than leaves from $NO_3^-$-depleted plants.

TABLE 1.   Nitrate depleted plants were transferred to
           Hoagland solution for different periods of
           time.  Leaves were detached from these plants
           and the changes in their nitrate and malate
           content determined two hours after detachment

| Hours in Hoagland | Nitrate reduced µeq/g fr. wt. | Malate accumulated µmoles/g fr. wt. |
|---|---|---|
| 0 | 0 | 0.1 |
| 3 | 2.8 | 2.3 |
| 6 | 3.5 | 3.1 |

In short term experiments the accumulation of
malate is similar to the amount of nitrate reduced
(Table 1).  The amount of malate accumulated during
the lifetime of the plant is much smaller than the
amount of reduced nitrogen found in plants.  One may
assume, therefore, that part of the malate formed in
response to nitrate reduction is consumed and does not
accumulate.  Dijkshoorn (1958) suggested that part of
the malate is translocated to the roots where it is
exchanged for nitrate.

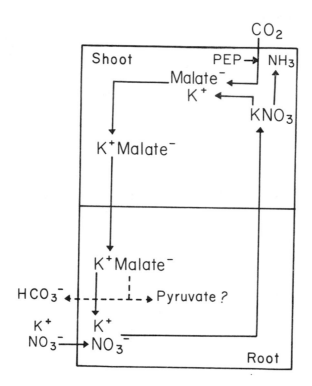

FIG. 1.  Model describing the movement of $K^+$, $NO_3^-$ and
         malate between shoot and root

This hypothesis is summarized in Fig. 1.  According-
ing to the scheme shown, $KNO_3$ is translocated to shoot
where nitrate is reduced and malate is synthesized con-
comitantly.  The $K^+$-malate formed is partially translocated to the root where malate is oxidized and $HCO_3^-$
exchanges for nitrate.  Nitrate taken up then moves
upwards together with the $K^+$ brought to the root,
through the phloem, with malate.

Table 2 shows that plants grown in half-strength
Hoagland solution (rich in nitrate) give up more labeled
$^{14}CO_2$ when transferred to Hoagland solution than to a

TABLE 2.   Plants grown in Hoagland solution were given
           a $^{14}CO_2$ pulse to one leaf.  Plants were then
           transferred to Hoagland solution (H–NO$_3$) or
           to a modified Hoagland solution in NO$_3^-$ was
           replaced by Cl$^-$ (H–Cl)

| Hours after $^{14}CO_2$ pulse | CPM $\times$ 10$^{-3}$ in culture solution | |
| | H – NO$_3$ | H – Cl |
|---|---|---|
| 0 | 0 | 0 |
| 5 | 4.3 | 8.3 |
| 22 | 36.9 | 13.6 |
| 70 | 34.1 | 15.9 |

modified Hoagland in which nitrate was replaced by Cl$^-$.
$^{14}CO_2$ was supplied under illumination to single leaves.
Rates of photo-synthesis were very similar.  Rates of
respiration of roots in Hoagland-NO$_3$ and Hoagland-Cl
were almost identical.  It is possible, therefore, that
the excess $^{14}CO_2$ given off in Hoagland is due to an
active exchange with nitrate.

According to the model proposed, roots of plants
actively reducing nitrate would be expected to take up
a large excess of nitrate over K$^+$.  The ratio of NO$_3^-$/K$^+$
uptake should decrease in nitrate depleted plants because malate is not synthesized in leaves.  Consequently, less malate is available for translocation to the
roots and for exchange with nitrate.  The data presented in Table 3 support these expectations.

From our model we also assumed that, no matter
how large the ratio of uptake NO$_3^-$/K$^+$, the ratio of
the ions in the exudate of the plants should always be
one, or close to one.  This assumption is based on the
fact that the exudate carries the ions moving upwards

TABLE 3.  Plants grown in Hoagland solution were
          transferred to 0.5 mM CaSO$_4$ for different
          periods of time.  After the indicated time
          in CaSO$_4$ plants were transferred to 1 mM
          solution of KNO$_3$ and the potassium and
          nitrate content of the solution was deter-
          mined after 90 minutes

| Days in CaSO$_4$ | Ion uptake: $\mu$eq/g fr. wt. | | |
| | NO$_3^-$ | K$^+$ | NO$_3^-$/K$^+$ |
| --- | --- | --- | --- |
| 1 | 25 | 4 | 6.25 |
| 5 | 14 | 15 | 0.93 |

from the root.  Nitrate taken up in excess of K$^+$ from
the external medium by exchange with bicarbonate will
accompany a corresponding amount of K$^+$ ions in the
root.  The external NO$_3^-$ will then move through the
transpiration stream together with K$^+$ brought down
from the shoot with malate.  This is shown in Table 4
where we can see that NO$_3$/K ratio is kept constant at
about 0.7.

The data shown here constitutes correlative evidence of the model proposed.  No direct evidence for
K$^+$ recirculation is as yet available.

The recirculation of K$^+$ ions in plants explains
their large requirement for K$^+$ ions in spite of the
fact that K$^+$ is not an organic constituent of any
structural compound.

The uptake of NO$_3$ by the plant is regulated to a
large extent by the metabolic utilization of the anion
in the shoot.  Plants reducing large amounts of nitrate
will have at their disposal more malic acid to facili-
tate the preferential uptake of nitrate.  Plants not

TABLE 4.  Experimental conditions as described in
          Table 3.  After 90 minutes uptake, plants
          were detopped, exudate collected for about
          2 hours and its nitrate and potassium
          content determined

| Days in CaSO₄ | Ion content of exudate : µeq/g fr. wt. | | |
|---|---|---|---|
| | $NO_3^-$ | $K^+$ | $NO_3^-/K^+$ |
| 1 | 5.8 | 8.6 | 0.67 |
| 5 | 3.1 | 4.3 | 0.72 |

needing nitrate (mature or flowering plants) will not
waste energy taking up large amounts of nitrate, which
is functionally useless to them.

   The proposed model allows a further explanation
of the curtailment of growth by NaCl at relatively
moderate concentrations.  $Cl^-$ interfers with nitrate
uptake and, therefore, with malate synthesis.  We have
observed that $Cl^-$ may replace $NO_3^-$ to some extent in
the exchange reaction with $HCO_3^-$ in the root.

   $Na^+$ may reach the shoot with $NO_3^-$ but Na-malate
will not move downwards to the root.  The immobiliza-
tion of malate in the shoot due to the presence of $Na^+$
will thus prevent the preferential uptake of $NO_3^-$ by
the roots.

   As a result of these effects of $Na^+$ and $Cl^-$ a
shortage of $NO_3^-$ will affect the plant, and growth
will be inhibited.

REFERENCES

1.  BAR-AKIVA, A. and SAGIV, J. (1967) Nitrate reduc-
    tase in the citrus plant:  properties, assay
    conditions and distribution within the plant.
    *Physiol. Plantarum 20*: 500-506.

2.  BEN-ZIONI, A., VAADIA, Y. and LIPS, S. HERMAN (1970)
    Correlations between nitrate reduction, protein
    synthesis and malate accumulation. *Physiol.
    Plantarum.*  (In press).

3.  DIJKSHOORN, W. (1958) Nitrate accumulation, nitrogen
    balance and cation-anion ratio during regrowth
    of perennial rye grass. *Ned. J. of Agric. Sci.
    6*: 211-221.

4.  SLOAN, C.H. and SUBLETT, B.J. (1966) Colorimetric
    method of analysis for nitrate in tobacco.
    *Tobacco Science X*: 121-125.

*Questions to Dr. Lips*

STEWARD:  Does it make any difference to your hypothesis
whether the reduction occurs here or there? You have to
circulate sugar.

LIPS:  There are two places of nitrate reduction: the
roots and the shoots.  You find a far larger activity of
the enzyme in the shoots than in the roots.  This, of
course, might be just a matter of adequate biochemical
essays.  But if you examine the exudate of plants you'll
find in plants with adequate nitrate nutrition that the
amount of nitrate nitrogen going up is far larger than
the amount of reduced nitrogen.

GUR:  There are plant species which reduce the nitrate
mainly in the roots. We know in case of apple plants,
that the extract of the leaves is sometimes more active
in the reduction of nitrates than root extracts. But
we almost don't find any nitrate exudates under conditions of normal nitrogen supply. How can you find a
connection between your theory and the behavior of
this species?

LIPS:  Our theory - and there are many people that have
been thinking in this direction - refers mainly to
herbaceous plants, and I'm quite sure that what you're
bringing up as an exception, is probably some tall tree.
I wouldn't extend it simply to the deciduous trees, because I'm quite sure that the mechanism is - at least
during spring - a different one.  I don't know what the
situation is during the summer or during the full growth
period of the tree; it might be similar then.

# INFLUENCE DU METABOLISME DE NO$_3^-$ DANS LES RACINES SUR L'ETAT NUTRITIONNEL DE LA PLANTE

## Yves Coïc

*Département de Physiologie végétale (I.N.R.A.), C.N.R.A., Versailles, France*

RÉSUMÉ

La puissance de métabolisme des nitrates (NO$_3^-$ → acide aminé) dans les racines est très différente suivant les espèces et variétés végétales. Les conséquences d'ordre nutritionnel de cet état de fait sont nombreuses et importantes. Selon nous:

Elle détermine la teneur cationique (K, Ca, Mg) et en acides organiques des feuilles des diverses espèces et règle l'alcalinisation physiologique du milieu nutritif.

Elle explique les différences du comportement des espèces vis à vis des antagonismes entre cations et, dans une certaine mesure l'état "calcicole" ou "calcifuge" des espèces végétales.

Elle joue un rôle considérable dans l'équilibre protidosynthèse-photosynthèse, et explique, par exemple, les possibilités *particulières* d'enrichissement en protides des grains de céréales par une alimentation azotée tardive.

Yves COÏC. Chef Dép. Physiol. Végét., C.N.R.A., Versailles, de 1956. né 1911 Gausson, Côtes du Nord (France); 1934 Ing. Agron., INA Pavi; 1941 Lic. et. Sci., Paris.

Elle explique l'effet différent, notamment sur la protidosynthèse, d'une déficience continue en un élément nutritif par rapport à une carence brutale survenant au cours de la végétation.

Elle permet d'expliquer ou de concevoir plus rationnellement certains tests nutritionnels ($NO_3^-$, $SO_4^=$ dans les organes de la partie aérienne).

## INTRODUCTION

Dans cet exposé le "métabolisme de $NO_3^-$" signifie le passage des nitrates à l'état d'acides aminés ou amides, c'est-à-dire le passage de l'azote nitrique à l'état d'azote organique. Il n'est donc pas limité à la première phase de réduction des nitrates par la nitrate-réductase.

On sait que le métabolisme des nitrates ainsi défini peut s'effectuer aussi bien dans les racines que dans les feuilles (et vraisemblablement dans d'autres organes), et qu'il y a entre les espèces végétales de grandes *différences d'aptitude* concernant la puissance de métabolisme des nitrates dans les racines.

Pour une même alimentation nitrique, la répartition du métabolisme des nitrates dans les racines et les feuilles sera fort différente suivant les genres de plantes.

Ce comportement différent des genres de plantes, relatif au métabolisme des nitrates, influe sur l'état nutritionnel de ces plantes et a, en conséquence, une incidence sur le choix du type d'analyse de la plante et l'interprétation des résultats.

Nous examinerons quelques unes des conséquences de la puissance plus ou moins grande du métabolisme des nitrates dans les racines.

INFLUENCE SUR LA TENEUR EN NITRATES, LA TENEUR
CATIONIQUE (K, Ca, Mg), LA TENEUR EN ACIDES
ORGANIQUES ET L'ALCALINISATION PHYSIOLOGIQUE

Lorsque l'ion nitrique migre vers les organes aé-
riens de la plante, il est associé aux cations K, Ca,
Mg, et la proportion de chacun d'entre eux qui est
transportée dépend essentiellement de la composition du
milieu alimentaire. Lorsque les nitrates sont trans-
formés en acides aminés ou amides dans la racine, ceux-
ci migrent (au pH de la sève ascendante) avec une faible
quantité de cations et il s'opère une sélection entre
les cations, le potassium migrant alors de façon privi-
légiée.

Le maïs et la tomate sont deux genres végétaux
ayant des aptitudes différentes concernant la puissance
de métabolisme des nitrates dans la racine, le maïs
ayant une puissance de métabolisme racinaire plus
grande que la tomate.

Pour une même alimentation en $NO_3^-$, il sera transporté plus de nitrates vers la feuille de tomate que
vers celle de maïs.

On peut en déduire quelques conséquences:

1. Le niveau auquel arrive le nitrate dans la
partie aérienne du maïs peut servir de test de l'abon-
dance de l'alimentation azotée nitrique d'une culture
de maïs (test à diphénylamine à l'intersection du limbe
et de la gaine foliaires).

2. Il a été constaté que lorsque l'alimentation
nitrique du pommier n'était pas très abondante, les

nitrates étaient entièrement métabolisés dans les fines
racines et qu'il ne montait pas de nitrates dans les
feuilles. On peut se demander quelle serait l'utilisation par les feuilles des nitrates apportés en pulvéri-
sation foliaire puisque l'on sait que la nitrate-réductase est une enzyme adaptive?

Lorsque les nitrates sont métabolisés dans la
feuille, l'équilibre électro-statique entre anions et
cations ($NO_3^-$ et $Ca^{++}$ par exemple) est maintenu grâce
aux acides organiques et il s'accumule dans la vacuole
des sels d'acides organiques. Le potassium peut se
redistribuer assez facilement (il est transporté notamment avec les acides aminés) mais la redistribution du
calcium est beaucoup plus difficile. En conséquence,
la teneur en acides organiques d'une feuille, mais
aussi celle en calcium, sont des tests de la quantité
de nitrates qui y ont été métabolisés. La teneur en
acides organiques des feuilles (notamment vieilles ou
adultes) n'est un indice de "santé" d'une plante que
parce qu'elle est un *témoignage* de la grandeur de la
nutrition azotée de la plante, processus physiologique
essentiel.

TABLEAU 1. Comparaison du maïs et de la tomate quant
aux teneurs en cations minéraux et en acides organiques en nutrition nitrique et
ammoniacale (en équivalents pour 100 g de
matière fraîche)

|  | Maïs | | Tomate | |
| --- | --- | --- | --- | --- |
|  | $NO_3$ | $NH_4$ | $NO_3$ | $NH_4$ |
| K | 13,1 | 13,1 | 11,2 | 8,3 |
| Ca | 6,2 | 4,0 | 26,2 | 4,2 |
| Mg | 5,2 | 3,3 | 9,5 | 3,0 |
| Acides organiques | 12,4 | 3,6 | 26,2 | 0,6 |

Le Tableau 1 montre la différence de composition des feuilles de tomate et de maïs en acides organiques et en cations minéraux bivalents (surtout en calcium) lorsque la nutrition est nitrique. Cette différence est beaucoup moins apparente lorsque la nutrition azotée est ammoniacale.

TABLEAU 2. Gains en acides organiques des feuilles de tabac pendant la période séparant deux prélèvements (en équivalent - mg par plante)

|  | - N - N | + N + N | - N + N |
|---|---|---|---|
| Acides organiques | 9,0 | 101,4 | 97,5 |

- N - N   Suppression de N ($NO_3^-$) de la solution nutritive 20 jours avant le 1er prélèvement (au premier prélèvement il restait encore un peu de nitrates dans la tige de tabac).

+ N + N   Plantes toujours bien alimentées en N ($NO_3^-$).

- N + N   Suppression de N, 20 jours avant le premier prélèvement et réapprovisionnement en N ($NO_3^-$) à partir du premier prélèvement.

Le Tableau 2 montre la différence de gains en acides organiques entre deux prélèvements chez les limbes foliaires de tabac lorsque la nutrition azotée a été, ou non, supprimée. Le Tableau 3 montre l'évolution au cours du temps de cette différence de gain pour les divers groupes de feuilles (feuilles jeunes, adultes, vieilles). Ce dernier tableau nous permet de suivre l'évolution de la puissance de métabolisme des nitrates dans les feuilles plus ou moins âgées au cours du temps.

Le tabac fait partie de la catégorie de plantes qui métabolisent une forte partie des nitrates dans les feuilles.

TABLEAU 3.  Evolution des différences de gains en aci-
            des organiques des divers groupes de limbes
            au cours du temps (en équivalent mg par
            plante)

|                | Fj   | Fa   | Fv   | Total |
|----------------|------|------|------|-------|
| du 20/7 au 9/8 | 14,2 | 31,2 | 35,6 | 81,0  |
| du 9/8 au 18/8 | 47,1 | 29,3 | 16,0 | 92,4  |

Fj : groupe de 6 feuilles "jeunes";
Fa : groupe de 4 feuilles "adultes";
Fv : groupe de 4 feuilles "vieilles".

Lorsque le métabolisme des nitrates se fait dans
les racines, l'équilibre électrostatique est assuré par
une absorption plus grande d'anions que de cations.  Il
y a *alcalinisation physiologique*.  Le Tableau 4 nous
montre la différence d'alcalinisation physiologique
chez le maïs et la tomate lorsque l'alimentation azotée
est approximativement la même.

TABLEAU 4

|                                                    | Maïs | Tomate |
|----------------------------------------------------|------|--------|
| pH final de la solution nutritive (pH initial = 6) | 7,3  | 6,6    |
| Alcalinisation (en méq.) / $NO_3^-$ absorbé (en méq.) | 0,20 | 0,04   |

DIFFERENCE DE COMPORTEMENT DES ESPÈCES VIS A VIS
DES ANTAGONISMES ENTRE CATIONS ET L'ETAT "CALCICOLE"
OU "CALCIFUGE" DES ESPÈCES VÉGÉTALES

Chez le maïs, le métabolisme de $NO_3^-$ dans la racine crée une barrière à la circulation du calcium et
aussi, mais moins forte, à celle du magnésium tandis que
cette barrière est faible dans le cas de la tomate.
Au niveau de la feuille, l'antagonisme du calcium vis-
à-vis du magnésium, par exemple, a beaucoup plus d'effet
chez la tomate que chez le maïs et l'on peut penser que
le maïs est moins affecté que la tomate par le calcaire
des sols. L'on comprend mieux le *déterminisme de l'é-
tat calcifuge* (tomate par rapport au maïs) des genres
végétaux.

IMPORTANCE DU METABOLISME DES RACINES DANS LA PROTIDO-
SYNTHÈSE - EQUILIBRE PHOTOSYNTHÈSE-PROTIDOSYNTHÈSE -
POSSIBILITES D'ENRICHISSEMENT EN AZOTE DES GRAINES DE
CÉRÉALES PAR L'ALIMENTATION AZOTÉE TARDIVE

La teneur en azote organique des plantes et fina-
lement des récoltes dépend essentiellement de l'allure
comparée de la protidosynthèse et de la photosynthèse.
La technique culturale a pour but d'obtenir la plus
forte photosynthèse nette à l'unité de surface de ter-
rain et la meilleure répartition et composition possible des produits de la photosynthèse. Les deux processus physiologiques, protidosynthèse et photosynthèse
ne sont évidemment pas indépendants.

L'importance de la racine dans la protidosynthèse
n'est pas assez soulignée. Du point de vue énergétique,

la phase la plus coûteuse dans la protéosynthèse est, de beaucoup, celle du passage de $NO_3^-$ aux acides aminés; celle de la phase ultérieure de synthèse des protéines étant relativement peu dépensière. En effet, le passage de $NO_3^-$ à $NH_4^+$ demande beaucoup d'énergie; puis il faut de l'énergie pour l'amination *réductrice* des acides $\alpha$ cétoniques; il faut aussi de l'énergie pour la synthèse de la fonction amide.

Donc, la synthèse des amino-acides dans la racine peut représenter dans certains genres végétaux une étape importante de la protéosynthèse. Chez les céréales, le blé par exemple, nous expliquons qu'il soit possible d'enrichir le grain par une alimentation azotée tardive, par le fait que le travail de synthèse des acides aminés dans la racine est encore très actif alors que la photosynthèse et la synthèse des acides aminés dans les feuilles diminuent fortement avec leur sénescence. On voit en effet le rapport anions/cations absorbés passer de 1,99 à 2,5 pour le blé tendre Florence Aurore et de 1,79 à 2,14 pour le blé dur Oued Zénati, sous l'influence de la nutrition azotée tardive ce qui, d'après nos conceptions, indique qu'une forte proportion des nitrates tardivement absorbés serait métabolisée dans la racine.

TABLEAU 5.  Influence de la nutrition azotée ($NO_3^-$) tardive (6 jours avant et 4 jours après floraison) du blé sur le rapport anions/cations de la plante (racines exceptées) à la récolte

|  | Florence Aurore blé tendre | Oued Zénati blé dur | Mahmoudi blé dur |
|---|---|---|---|
| sans azote tardif | 1,99 | 1,79 | 1,87 |
| avec azote tardif | 2,50 | 2,14 | 2,14 |

EFFET DU MODE DE DEFICIENCE EN UN ELEMENT
SUR LA NUTRITION AZOTEE DE LA PLANTE

Nous avons été amenés à travailler sur deux moda-
lités de déficience en un élément, P par exemple:

1. Une alimentation en P restreinte mais continue: c'est ce qui se passe dans un sol pauvre en P
assimilable;

2. Une alimentation en P normale suivie d'une
suppression complète en cet élément. C'est ce que
l'on pratique souvent lors d'expériences en solutions
nutritives, ce qui permet de disposer d'une quantité
suffisante de matériel végétal à analyser et d'obtenir
des symptômes rapides de déficience causée par le phénomène de dilution du P par la nouvelle matière végétale formée.

Dans le premier cas, c'est la partie aérienne de
la plante qui souffre plus que la partie racinaire qui
utilise le peu de phosphore qu'elle trouve. En conséquence le rapport racines/parties aériennes est plus
grand que pour une plante normale et surtout, le rapport protidosynthèse/photosynthèse s'accroit car, dans
la feuille, la photosynthèse comme la protidosynthèse
sont très déficientes alors que, dans la racine, la
protidosynthèse est moins diminuée. Le grain des céréales dans une parcelle déficiente en P sera plus
riche en azote que celui des parcelles enrichies en P.

Dans le deuxième cas (suppression de P dans la
solution nutritive), la protidosynthèse globale est
plus affectée que la photosynthèse parce que la protidosynthèse dans la racine est fort touchée. Alors,
la plante déficiente en P est aussi déficiente en N
organique par rapport à la plante normalement alimen-

tée: la teneur de limbe du maïs passe de 4,63% de la
matière sèche lorsque l'alimentation en P est normale,
à 3,42% lorsque le P a été supprimé totalement de la
solution nutritive.

## AUTRES EFFETS POSSIBLES

Dans certaines de nos études sur la sensibilité
de différents genres végétaux à la simazine, il nous
semble qu'il y ait une corrélation entre la sensibilité
à cette triazine et la puissance de métabolisme des nitrates dans les racines. La résistance à la simazine
tiendrait à la transformation dans la racine de la
simazine en hydroxysimazine, dérivé qui ne manifeste
pas de toxicité. Le maïs fait partie des plantes ré-
sistantes et la tomate des plantes sensibles.

## INFLUENCE DES FACTEURS POUVANT FAIRE VARIER
## LA PART PLUS OU MOINS GRANDE PRISE PAR LA RACINE
## DANS LA PROTIDOSYNTHESE

En agriculture, la connaissance de l'influence des
divers facteurs pouvant faire varier la part prise par
la racine dans la protidosynthèse est très importante.
Un facteur primordial est évidemment la température et,
en Agriculture, la température des racines est très
souvent différente de celle de la partie aérienne, notamment dans les régions tempérées au début du prin-
temps, le sol se réchauffant moins vite que l'air.

Nous avons parlé du mode de déficience en un élément (P). La transpiration est aussi un facteur ayant
une influence puisque, suivant son intensité, elle
modifie la vitesse de transport des nitrates dans la
feuille. Des variations nycthémérales doivent exister

quant à la part prise par la racine dans la protidosynthèse et ces variations peuvent avoir plus d'ampli-
tude selon les espèces végétales. Evidemment, la
grandeur de l'alimentation nitrique influe aussi.

CONCLUSIONS

La part plus ou moins grande que prend la racine
dans le métabolisme de $NO_3^-$ a, ainsi que nous l'avons
exposé, une grande influence sur l'état nutritionnel
des végétaux et, en conséquence, une importance considérable en Agronomie, qu'il s'agisse de l'amélioration
du matériel végétal, de son fonctionnement, des déterminations analytiques à effectuer pour mieux diagnosti-
quer l'état alimentaire du milieu ou ce qu'a été le
fonctionnement de la plante, d'établir la nutrition
optimum pour chaque espèce végétale.

# INFLUENCE OF VARIOUS FACTORS ON NITRATE CONCENTRATION IN PLANTS IN RELATION TO NITROGEN METABOLISM

Chresten Sørensen

*Department of Agricultural Chemistry,
State Laboratory for Soil and Crop Research,
Lyngby, Denmark*

ABSTRACT

The concentration of nitrate in plants is significant for plant physiology, fertilization and animal nutrition. Field and pot experiments with ryegrass, oats, barley, wheat, beets and red clover have shown that the concentration of nitrate varies considerably from one plant organ to another and from one beet-top petiole to another within the same plant.

Following fertilization with high amounts of nitrate, the concentration of nitrate in the plant can be increased even further by additional application of K. Addition of P, Cl or $SO_4$ decreases the concentration to a small extent.

It is found that nitrate activates nitrate reductase, but, after fertilization with high amounts of nitrogen, the nitrate concentration in the plant can rise to toxic levels.

Nitrate has been found to be the most variable of the nitrogen fractions studied.

Chresten SØRENSEN. Head Dept. Agric. Chem., State Lab. Soil & Crop Res., Lyngby (Denmark) since 1963.
 b. 1918 Hjorning (Denmark); 1946 B.Sc. and 1954 Ph.D., Roy. Vet. Agric. Coll., Copenhagen; 1954 Res.
 Assoc., Roy. Vet. Agric. Coll., Copenhagen, Denmark.

Nitrogen metabolism in plants is of interest from various points of view, one of which is obtaining better understanding of protein production. A great amount of the nitrogen taken up by the non-leguminous plants is absorbed as nitrate nitrogen and further bound into organic components.

The concentration of nitrate in plants gives some idea of the reducing capacity of the plant, and as the processes involved make the first main step for the further production of protein, knowledge of which factors influence the nitrate concentration is of great importance.

It is well known that enzymes involved in the processes are activated by the substrate (Beevers *et al.*, 1965), molybdenum is necessary (Afridi and Hewitt, 1964), and other growth factors, such as climate and plant hormones, have an influence on the nitrate reduction and thereby on the remaining concentration.

In order to obtain information concerning the effect of some few nutrients on the concentration of nitrate in various plants and to get an idea of the importance of the process in relation to other main steps in the nitrogen metabolism, the following experiments have beeen carried out.


METHODS


*Experiment A*:   In order to investigate the rate of nitrate absorption under field conditions and the following variation in contents of nitrogen fractions, a sward of foxtail (*Dactylis glomerata*) was harvested on August 2. The experimental plot was supplied with 50 g of nitrate nitrogen as calcium nitrate followed by 10 litres of water per square meter.

The grass was harvested from small subplots at different stages of development during a six weeks' period, at the end of which the grass had reached a height of 60 cm.  The crop was harvested by scissors in order to avoid contamination from older stubbles.

_Experiment B_ was carried out to obtain information of the magnitude of the influence of chloride and sulphate under field conditions.  A sward of ryegrass (_Lolium perenne_) growing on a fertile soil was supplied with 250 kg of K, 90 kg of P and 186 kg of N (as calcium nitrate) per ha.  As seen from Table 2, the plots were fertilized with equal amounts of chlorides or sulphates as potassium or calcium salts.

_Experiment C_ was performed in order to investigate the influence of nitrate fertilization on nitrate content in red clover (_Trifolium pratense_). A sward of this plant on a fertile soil was supplied with increasing amounts of calcium nitrate, and three successive cuts were taken.

_Experiment D_ was carried out as a pot experiment using plastic pots containing 24 kg of a mixture consisting of one-third poor soil with two-thirds silver sand.  The surface area of the pots was 500 square cm.  The nutrients were mixed with the growth medium, seeds of ryegrass (_Lolium perenne_) were sown on April 16, and five replicates were harvested on June 12.  The growth medium was given a basic dressing of micro nutrients.  Some of the treatments are shown in Table 4, and grams of nutrients per pot were as follows:

$$N_0 = 0 \qquad K_0 = 0 \qquad P_0 = 0$$
$$N_1 = 0.75 \qquad K_1 = 2 \qquad P_1 = 1$$
$$N_2 = 1.5 \qquad K_2 = 8 \qquad P_2 = 4$$
$$N_3 = 3.0$$
$$N_4 = 6.0$$
$$N_5 = 9.0$$

The growth medium was supplied daily with up to 80% of its water-holding capacity, calculated on dry matter basis.

Determination of dry matter (DM) was carried out at 80°C for 24 hours. Total nitrogen (TN) was determined by the Kjeldahl method using salicylic acid, and the selenium catalyst (Chibnall *et al.*, 1943). Protein nitrogen (PN) was determined by the Stutzer method (Analysemetoder, 1958). Nitrate nitrogen ($NO_3$-N) was determined by a modification of the xylenol method (Sørensen, 1956). Soluble nitrogen (SN) is calculated as the difference between TN and PN. Soluble organic nitrogen (SON) is calculated as the difference between SN and $NO_3$-N.

RESULTS

*Experiment A*: Harvest was carried out every third or fourth day, and representative results are shown in Table 1.

In the heavy nitrogen fertilized regrowth, the percentage of TN increases for the first 10 days, thereafter the nitrogen content is almost constant for about twenty days, after which it slowly decreases.

The PN as a percentage of TN varies during most of the growth period to a rather small extent, which probably may be due to a high capacity of protein synthesis in the young tillers. The results from the shoots taken on the third day make an exception which is seen from the ratio PN/SON. Furthermore the relative content of SON at the third day is high and so is the ratio SON/$NO_3$-N. The results show that protein synthesis is delayed compared with the absorption and reduction of nitrate. After this first step, the relative content of $NO_3$-N still increases, but SON as a percentage of TN decreases, as does the ratio SON/$NO_3$-N, which means that reduction of nitrate has decreased relatively. However,

the ratio PN/SON has increased, and during the main part
of the growth period the discrepancy in protein synthesis
seems to be a lack of nitrate reduction.

At the end of the period, the response to the decrease in TN and $NO_3$-N in the plant is a relative increase in protein synthesis, which is seen from the
ratio PN/SON, whereas the ratio SON/$NO_3$-N varies only
to a small extent.  This means that the nitrate reduction is still insufficient.

*Experiment B*:  The results from experiment B are
seen in Table 2.  Fertilization with chloride and
sulphate has repressed the concentration of TN and the
content of $NO_3$-N as a percentage of TN.  The tendencies
in the figures for the ratio PN/SON are that the ratio
is smaller for the control than for the treated plants,
which means that so far protein synthesis has increased.

*Experiment C* was carried out with red clovers and
at the first cut the content of TN increased from 3.8
to 5.2 percent of DM, when fertilization with nitrate
increased from zero up to 600 kg of N  per ha, and the
percentage of $NO_3$-N varied from 0.02 to 0.2, respectively.  The results from the second cut are seen in
Table 3.

The content of TN for the first two additions of
fertilizers are smaller than or of the same magnitude
as the control, but for the higher nitrogen fertilization, a correspondingly higher percentage of TN is
found.

The percentage of $NO_3$-N has increased with increasing fertilization.  The ratio PN/SON differs considerably only for the application of 600 kg of nitrogen per ha, which may be a result of decreasing protein synthesis.  But SON/$NO_3$-N decreases appreciably,
which may be related to insufficient reduction of
nitrate.

TABLE 1

Percentage of Total Nitrogen and Nitrogen Fractions in Foxtail
(Experiment A)

| Date of Harvest | g DM per plot | pCt TN | % of TN PN | % of TN NO$_3$-N | SON | SON/NO$_3$-N | PN/SON |
|---|---|---|---|---|---|---|---|
| 2/8 | -- | 1.6 | 94 | 0.04 | 4 | 97.5 | 24 |
| 5/8 | 2 | 3.2 | 79 | 5 | 16 | 3.2 | 5 |
| 8/8 | 6 | 3.6 | 81 | 10 | 9 | 0.9 | 9 |
| 13/8 | 15 | 4.5 | 74 | 16 | 11 | 0.7 | 7 |
| 22/8 | 35 | 4.6 | 76 | 14 | 10 | 0.7 | 8 |
| 1/9 | 87 | 4.0 | 83 | 11 | 7 | 0.6 | 13 |
| 14/9 | 112 | 3.6 | 84 | 10 | 7 | 0.7 | 13 |

TABLE 2

Nitrogen Content and Nitrogen Fractions in Ryegrass
(Experiment B)

| Kg per ha | kg DM per plot | pCt TN | Percentage of TN | | | SON/$NO_3$-N | PN/SON |
|---|---|---|---|---|---|---|---|
| | | | PN | $NO_3$-N | SON | | |
| 1000 $K_2SO_4$ | 7.1 | 3.4 | 77 | 10 | 14 | 1.4 | 6 |
| 1000 K Cl | 7.1 | 3.6 | 75 | 9 | 17 | 1.9 | 5 |
| Control | 6.4 | 4.0 | 71 | 13 | 17 | 1.3 | 4 |
| 666 $CaCl_2$ | 6.4 | 3.7 | 77 | 8 | 15 | 2.0 | 5 |
| 785 $CaSO_4$ | 6.6 | 4.0 | 75 | 11 | 14 | 1.3 | 5 |

TABLE 3

Percentage of Total Nitrogen and Nitrogen Fractions at Increasing Nitrate Fertilization of Red Clover
(Experiment C)

| Kg N per ha | g DM per plot | pCt TN | Percentage of TN | | | SON/NO$_3$-N | PN/SON |
|---|---|---|---|---|---|---|---|
| | | | PN | NO$_3$-N | SON | | |
| 0 | 101 | 3.4 | 86 | 1 | 14 | 18.8 | 6 |
| 75 | 140 | 3.1 | 83 | 2 | 15 | 7.3 | 6 |
| 150 | 142 | 3.4 | 83 | 4 | 14 | 3.8 | 6 |
| 300 | 174 | 3.7 | 78 | 6 | 15 | 2.4 | 5 |
| 600 | 147 | 3.7 | 73 | 8 | 19 | 2.3 | 4 |

*Experiment D*:   Some of the results from this pot
experiment with ryegrass are presented in Table 4.
The figures show that at low nitrogen level, $N_1$, the
addition of potassium produces an increase in dry
matter yield, followed by a fall in the percentage
of TN, and the relative amounts of various nitrogen
fractions are almost unchanged.  At the highest
nitrogen level, note the relative smaller content of
PN and high content of $NO_3$-N, resulting in an almost
unchanged order of magnitude of SON as a percentage of
TN.  However, the relative content of SON for the two
series with high nitrogen level decreases with increasing fertilization with potassium.  These results,
evaluated in connection with the great differences in
the content of $NO_3$-N, show that with increasing potassium fertilization the protein synthesis is enhanced to a small extent, and the most varied nitrogen fraction is $NO_3$-N, which increases at high nitrogen levels, but is almost unaffected at low levels.

At increasing fertilization with P, the production
of dry matter increases appreciably, followed by a
sharp decrease in the content of TN at the $N_1K_1$ level,
but at the high N level the percentage of TN increases
to a comparatively small extent.  For the low N level,
$N_1$, the absorption of nitrogen is 270 mg of N for $P_0$,
but for the $P_1$ fertilization, the absorption is 490 mg
of N.  Therefore, some of the effect of P might be
metabolic and not only a result of dilution.  Also
the relative content of PN and SON varies for fertilization with phosphorus, but in opposite direction at
the two levels of N.  At the $N_1$ level, the addition of
P has accelerated the reduction of nitrate appreciably,
which may be a function of enhanced photosynthesis,
following a higher concentration of phosphorus and
ATP in the plant.  However, the protein synthesis does
not follow the nitrate reduction, as seen from the
figures for PN/SON, which has decreased.

At the high nitrogen level, the addition of phos-

TABLE 4
Dry Matter Yield, Nitrogen Content and Nitrogen Fractions in Ryegrass (Experiment D)

| Treatment | g DM per plot | pCt TN | Percentage of TN | | | SON/NO$_3$-N | PN/SON |
|---|---|---|---|---|---|---|---|
| | | | PN | NO$_3$-N | SON | | |
| N$_0$P$_1$K$_1$ | 2.7 | 1.9 | 92 | 3 | 6 | 2.2 | 16 |
| N$_1$P$_1$K$_1$ | 15.3 | 3.2 | 92 | 1 | 7 | 8.1 | 13 |
| N$_2$P$_1$K$_1$ | 19.9 | 4.9 | 83 | 9 | 8 | 0.9 | 10 |
| N$_1$P$_1$K$_0$ | 13.5 | 3.8 | 90 | 2 | 8 | 3.9 | 11 |
| N$_1$P$_1$K$_1$ | 15.3 | 3.2 | 92 | 1 | 7 | 8.1 | 13 |
| N$_1$P$_1$K$_2$ | 16.1 | 3.2 | 91 | 1 | 8 | 5.8 | 12 |
| N$_4$P$_1$K$_0$ | 13.7 | 5.4 | 79 | 13 | 9 | 0.7 | 9 |
| N$_4$P$_1$K$_1$ | 20.5 | 5.7 | 77 | 16 | 7 | 0.4 | 11 |
| N$_4$P$_1$K$_2$ | 21.1 | 5.6 | 77 | 17 | 6 | 0.3 | 13 |
| N$_3$P$_2$K$_0$ | 17.3 | 5.4 | 80 | 12 | 8 | 0.7 | 10 |
| N$_3$P$_2$K$_1$ | 23.7 | 5.7 | 77 | 15 | 8 | 0.5 | 10 |
| N$_3$P$_2$K$_2$ | 26.1 | 5.6 | 77 | 17 | 5 | 0.3 | 15 |
| N$_1$P$_0$K$_1$ | 5.1 | 5.3 | 82 | 14 | 3 | 0.2 | 25 |
| N$_1$P$_1$K$_1$ | 15.3 | 3.2 | 92 | 1 | 7 | 8.1 | 13 |
| N$_1$P$_2$K$_1$ | 16.5 | 3.1 | 92 | 1 | 7 | 7.2 | 13 |
| N$_4$P$_0$K$_1$ | 3.3 | 5.1 | 79 | 11 | 10 | 0.9 | 8 |
| N$_4$P$_1$K$_1$ | 20.5 | 5.7 | 77 | 16 | 7 | 0.4 | 11 |
| N$_4$P$_2$K$_1$ | 22.0 | 5.7 | 76 | 17 | 7 | 0.4 | 11 |

phorus is followed by a certain increase in the nitrate content and a small decrease in SON, which results in a relative increase in protein synthesis evaluated by the ratio PN/SON. Although the variation in concentration of $NO_3$-N is comparatively small in this part of the experiment, it is the most variable fraction.

SUMMARY


    Experiments with ryegrass and red clover have been carried out in the field and in pots as well. The concentration of nitrate varies considerably. The relations to concentrations of total nitrogen, protein nitrogen and soluble organic nitrogen are calculated. Fertilization with the nutrients N, P, K, Cl, and S influences the nitrate concentration. After fertilization with high amounts of nitrate the concentration of nitrate in the plant increases further by increasing fertilization with potassium. Applications of chlorides or sulphates decrease the concentration to a small extent. After application of phosphate at low nitrate fertilization, the nitrate concentration decreases.

    After harvest and nitrate fertilization of ryegrass, the nitrate concentration increased during the first ten days of regrowth. The nitrate concentration in plants seems to reach an upper level, when increasing amounts of nitrate are added to the growth medium. Nitrate has been the most variable of the nitrogen fractions studied.


REFERENCES


1.  AFRIDI, M.M.R.K. and HEWITT, E.J. (1964)  *J. Exptl.*

*Botany 15:* 251-271.

2.   *Arbejdsmetoder for Kemiske Undersøgelser af Maelk
     og Mejeriprodukter m.m.*   København, 1941.

3.   BEEVERS, L., SCHRADER, L.E., FLESHER, DONNA and
     HAGEMAN, R.H. (1965)  *Plant Physiol. 40:* 691-698.

4.   CHIBNALL, A.C., REES, M.W. and WILLIAMS, E.F. (1943)
     *Biochem. J. 37:* 354-359.

5.   VICKERY, H.B., PUCHER, G.W., CLARK, H.E., CHIBNALL,
     A.C. and WESTALL, R.G. (1935)  *Biochem. J. 29:*
     2710-2720.

6.   SØRENSEN, C. (1956)  *Physiol. Plant. 9:* 304-320.

# SOLUBLE NITROGENOUS FRACTIONS OF TISSUE EXTRACTS AS INDICES OF THE NITROGEN STATUS OF PEACH TREES*

Brian K. Taylor **

*Department of Agriculture, Horticultural Research Station, Tatura, Victoria Australia*

ABSTRACT


We show that soluble nitrogenous fractions of extracts of tissues of dormant peach trees are sensitive indices of the nitrogen status of both young and mature trees.  Indices of the nitrogen status of mature trees based on levels of soluble nitrogenous fractions in root and bud tissues were more sensitive than conventional leaf analysis for total N in midshoot leaves in mid summer.  Of the various fractions measured, it is suggested that routine tests for assessing the nitrogen status of dormant trees should be based on measurement of levels of total alpha-amino N or soluble N in tissue extracts. The main advantage of assessing nitrogen status in this way is that we may make an estimate on dormant trees prior to commencement of growth in spring.

---

\* Read at the Colloquium by R.S. Harper
\*\*Present address:  Department of Agriculture, Scoresby Horticultural Research Station, P.O. Box 174, Ferntree Gully, Victoria, Australia, 3156

## INTRODUCTION

It is universally recognized that annual applications of nitrogenous fertilizer need to be applied to
fruit trees in most fruit growing areas of the world in
order to maintain growth and yield potential. Indeed,
on some fertile soils in semi-arid regions nitrogen has
been the only fertilizer element necessary for producing
high yields and consistent bearing in the long-term
(Ballinger, 1965). As a consequence, considerable
effort has been employed on a world-wide scale to
develop leaf and other forms of plant analysis for
assessment of the nitrogen status of fruit trees (e.g.
Oland, 1959; Baxter 1965; Bould, 1966).

In this paper I wish to present data to show that
soluble nitrogenous fractions of extracts of tissues of
dormant trees are sensitive indices of the nitrogen
status of peach trees. Development of this approach
arose from the classical work of Oland (1954; 1959) on
the nitrogenous reserves of apple trees.

## ANALYTICAL PROCEDURES

Various tissues were harvested from young and
mature peach trees receiving differential levels of
nitrogenous fertilizer and analyzed as follows:

1. *Woody tissue and bud analyses.* Finely ground,
freeze-dried tissues from dormant trees were exhaus-
tively extracted with 0.05 $M$ citrate buffer, pH 5.0 or
other solvents and levels of soluble N, arginine N,
total alpha-amino N, amide N, ammonium N and nitrate N
were measured as described elsewhere (Taylor and May,
1967; Taylor and van den Ende, 1969). Total N levels
in dried tissues were also measured (Kjeldahl analysis)
and hence insoluble N values could be calculated

by difference.

2. *Leaf analyses.* Health, midshoot leaves from current shoots were ovendried and analyzed for total N content by Kjeldahl analysis. Abscissed leaves were collected as described by Taylor and van den Ende (1969).


RESULTS


A. *Young Trees*

It is evident from Table 1 that the concentration of free arginine in woody tops (minus buds) was the most sensitive index of the nitrogen status of the young trees. Generally, concentrations of organic nitrogenous fractions (soluble N, arginine N, total alpha-amino N and amide N) were excellent indices of nitrogen status in comparison with total N, insoluble N and ammonium N levels, and similar results were obtained if the data were expressed on an absolute basis as reported else-where (Taylor and May, 1967). Concentrations of ammonium N were very low in all tissues and no nitrate N was detected.


B. *Mature Trees*

In contrast to the situation with young peach trees in sand culture, the data in Table 2 show that the concentration of arginine N in the roots was the most sensitive index of the nitrogen status of mature trees in the field. It is probable that this difference was due to differences in the nature of the roots and the proportion of the total root available for analysis; in the case of the young trees the total root system was subsampled for analysis whereas with the trees in the field root pieces (0.5 to 2.0 cm thick plus associated fibrous

TABLE 1

Indices of the Nitrogen Status of Two-year-old Peach Trees Grown in Sand Culture (Values are percentages relative to concentrations in $N_1$ tissues; trees were harvested in mid July after receiving differential nitrogen treatments for one year).

| Tree Tissue | Relative Nitrogen Level* | Total N | Insol- uble N | Soluble N | Arginine N | Total Alpha- amino N | Amide N | Ammonium N |
|---|---|---|---|---|---|---|---|---|
| Roots | $N_1$ | 100.0 | 100.0 | 100.0 | 100.0 | 100.0 | 100.0 | 100.0 |
| | $N_3$ | 134.1 | 110.3 | 187.6 | 195.7 | 193.4 | 136.4 | 134.4 |
| | $N_9$ | 173.4 | 106.3 | 324.2 | 384.8 | 304.6 | 155.8 | 121.9 |
| Tops (minus buds) | $N_1$ | 100.0 | 100.0 | 100.0 | 100.0 | 100.0 | 100.0 | 100.0 |
| | $N_3$ | 152.7 | 118.4 | 258.1 | 284.0 | 276.3 | 209.7 | 107.0 |
| | $N_9$ | 186.1 | 128.6 | 362.5 | 480.4 | 379.3 | 237.3 | 77.2 |
| Leaf + flower buds | $N_1$ | 100.0 | 100.0 | 100.0 | 100.0 | 100.0 | 100.0 | 100.0 |
| | $N_3$ | 127.3 | 105.2 | 240.6 | 328.6 | 225.4 | 241.4 | 93.7 |
| | $N_9$ | 154.7 | 120.5 | 331.0 | 452.3 | 301.8 | 251.4 | 81.1 |

* Nitrogen was supplied as nitrate.

TABLE 2

Indices of the Nitrogen Status of Mature Peach Trees in the Field (values are percentages relative to concentration in No tissues)
(a) *Dormant Tree Analyses* Tissues were harvested in mid July from trees which had recieved differential nitrogen treatments for one growing season).

| Tree Tissue | Relative Nitrogen Level* | Total N | Insoluble N | Soluble N | Arginine N | Total Alpha-amino N | Amide N | Ammonium N |
|---|---|---|---|---|---|---|---|---|
| Roots | $N_0$ | 100.0 | 100.0 | 100.0 | 100.0 | 100.0 | 100.0 | 100.0 |
| | $N_3$ | 134.6 | 130.2 | 139.6 | 167.2 | 187.7 | 169.0 | 53.8 |
| | $N_6$ | 131.5 | 102.9 | 164.0 | 184.7 | 223.3 | 193.1 | 38.5 |
| | $N_{12}$ | 157.6 | 122.4 | 198.0 | 251.9 | 248.3 | 189.7 | 7.7 |
| | $N_{18}$ | 153.6 | 123.0 | 188.4 | 271.8 | 228.3 | 141.4 | 38.5 |
| 2-3-year old shoots | $N_0$ | 100.0 | 100.0 | 100.0 | 100.0 | 100.0 | 100.0 | 100.0 |
| | $N_3$ | 95.8 | 96.2 | 94.5 | 98.7 | 97.1 | 84.4 | 50.0 |
| | $N_6$ | 91.9 | 91.7 | 92.4 | 82.7 | 91.2 | 46.7 | 275.0 |
| | $N_{12}$ | 99.7 | 93.7 | 121.4 | 140.0 | 126.5 | 35.6 | 287.5 |
| | $N_{18}$ | 108.1 | 101.7 | 131.0 | 161.3 | 135.3 | 60.0 | 625.0 |
| Current shoots (minus buds) | $N_0$ | 100.0 | 100.0 | 100.0 | 100.0 | 100.0 | 100.0 | 100.0 |
| | $N_3$ | 105.6 | 103.5 | 113.7 | 110.7 | 112.0 | 158.1 | 100.0 |
| | $N_6$ | 100.9 | 98.4 | 109.6 | 95.2 | 108.0 | 100.9 | 57.1 |
| | $N_{12}$ | 110.8 | 104.0 | 132.9 | 148.8 | 130.0 | 137.6 | 71.4 |
| | $N_{18}$ | 114.9 | 103.5 | 154.3 | 163.1 | 142.0 | 110.3 | 71.4 |
| Leaf + flower buds | $N_0$ | 100.0 | 100.0 | 100.0 | 100.0 | 100.0 | 100.0 | 100.0 |
| | $N_3$ | 100.7 | 101.0 | 98.5 | 94.6 | 96.2 | 94.6 | 55.6 |
| | $N_6$ | 107.2 | 105.4 | 122.1 | 122.8 | 107.7 | 89.9 | 66.7 |
| | $N_{12}$ | 113.0 | 108.6 | 140.7 | 164.1 | 151.9 | 68.2 | 133.3 |
| | $N_{18}$ | 117.4 | 109.2 | 166.8 | 195.7 | 180.8 | 47.3 | 181.5 |

* Nitrogen was supplied as nitrochalk fertilizer ($CaCa_3$ + $NH_4NO_3$, 20.5 % N). Half was applied per tree square in mid December and the other half was similarly applied in late March.

TABLE 2

(b) *Leaf Analysis* (Leaves were harvested on several occasions through a growing season from the trees mentioned in part (a) above; differential nitrogen treatments were continued).

| Index | Relative Nitrogen Treatment | Date Leaves Harvested | | | | | |
|---|---|---|---|---|---|---|---|
| | | Oct. 31 | Dec. 11 | Jan. 22 | Mar. 20 | Apr. 23 | Abscissed Leaves |
| Concn. total N in leaves | $N_0$ | 100.0 | 100.0 | 100.0 | 100.0 | 100.0 | 100.0 |
| | $N_3$ | 116.7 | 122.4 | 112.9 | 116.6 | 127.5 | 126.2 |
| | $N_6$ | 118.1 | 118.6 | 118.6 | 122.0 | 126.8 | 150.6 |
| | $N_{12}$ | 124.5 | 138.2 | 143.1 | 137.8 | 140.7 | 168.8 |
| | $N_{18}$ | 125.0 | 141.9 | 150.3 | 154.8 | 156.8 | 172.1 |
| Amount of total N per leaf | $N_0$ | 100.0 | 100.0 | 100.0 | 100.0 | 100.0 | 100.0 |
| | $N_3$ | 131.5 | 138.4 | 119.0 | 136.7 | 144.6 | 133.0 |
| | $N_6$ | 123.8 | 122.8 | 126.3 | 135.1 | 134.7 | 155.5 |
| | $N_{12}$ | 140.1 | 150.0 | 161.7 | 161.2 | 157.9 | 181.9 |
| | $N_{18}$ | 142.0 | 160.0 | 172.0 | 187.1 | 180.0 | 180.8 |

roots) were sampled for analysis.  Since there could be
objections to using root tissue from mature trees for
analysis, it is of interest to note that the next best
estimate of the nitrogen status of the trees was given
by the concentration of arginine N in leaf + flower
buds.

Arginine levels in roots of dormant trees gave a
more sensitive estimate of tree nitrogen status than
levels of total N in leaves at any time during the growing season, irrespective of whether leaf analysis data
were expressed on a concentration or absolute basis.
Generally, as for the young trees, levels of organic
nitrogenous fractions in extracts of woody tissues of
dormant trees were excellent indices of nitrogen status.


DISCUSSION


Results presented in this paper are in agreement
with those of Baxter (1965) and show that concentrations of soluble N, arginine N and total alpha-amino
N in tissues of dormant peach trees are highly sensitive indices of the nitrogen status of the trees.  A
similar situation is known to hold for apple (Oland,
1959; Baxter, 1965), citrus (Baxter, 1965) and pear
(Taylor, unpublished data) trees and is also likely to
hold for other fruit tree species as well, especially
those belonging to the Family Rosaceae since free arginine has been shown to be an important nitrogenous
storage compound in many species of this family (Reuter,
1957; Oland, 1959; Baxter, 1965; Taylor, 1967).

Generally, estimates of the nitrogen status of dormant peach trees were more sensitive if based on arginine N levels than if based on levels of total alpha-
amino N or soluble N.  Routine measurement of free arginine levels in crude extracts of plant tissues is
usually based on the Sakaguchi reaction but flavonoid

materials interfere in color development (Taylor and May, 1967; Taylor and van den Ende, 1969) and the reaction color rapidly fades (Gilboe and Williams, 1956). However, in contrast, analysis for total alpha-amino N and soluble N is straight forward and reliable (see Taylor and May, 1967) and either fraction is suitable as a basis for the routine assessment of the nitrogen status of dormant fruit trees. Flavonoid materials do not interfere to any significant extent in the measurement of total alpha-amino N levels in peach extracts.

Although estimates of the nitrogen status of peach trees based on levels of soluble nitrogenous fractions in root and bud tissues tended to be more sensitive than those obtained from conventional leaf analysis for total N, it is suggested that the practical use of analysis for such fractions lies in the fact that they allow an estimate to be made of the nitrogen status of deciduous trees prior to commencement of growth in spring. That is, it is now possible to determine tissue standards for the nitrogen status of orchard trees in winter as well as in mid summer.

REFERENCES

1.    BALLINGER, W.E. (1965) Peach fertilization. *Proc. Peach Congress*, Verona, Italy, pp. 405-439.

2.    BAXTER, P. (1965) A simple and rapid test, using the ninhydrin method, for the determination of the nitrogen status of fruit trees. *J. Hort. Sci. 40*: 1-12.

3.    BOULD, C. (1966) Leaf analysis of deciduous crops. Childers, N.F. (ed.) *Fruit Nutrition 2nd ed.*: 651-684. Somerset Press, Somerville.

4.    GILBOE, D.D. and WILLIAMS, J.W. (1956) Evaluation of the Sakaguchi reaction for the quantitative determination of arginine. *Proc. Soc. Exp. Biol. Med. 91*: 535-536.

5.    OLAND, K. (1954) Nitrogenous constituents of apple maidens grown under different nitrogen treatments. *Physiol. Plant. 7*: 463-474.

6.    ————— (1959) Nitrogenous reserves of apple trees. *Physiol. Plant. 12*: 594-648.

7.    REUTER, G. (1957) Die Hauptformen des löslichen Stickstoffs in vegetativen pflanzlichen Speicherorganen und ihre systematische Bewertbarkeit. *Flora. 145*: 326-338.

8.    TAYLOR, B.K. (1967) Storage and mobilization of nitrogen in fruit trees: a review. *J. Aust. Inst. Agric. Sci. 33*: 23-29.

9.    TAYLOR, B.K. and MAY, L.H. (1967) The nitrogen nutrition of the peach tree II.  Storage and mobilization of nitrogen in young trees. *Aust. J. Biol. Sci. 20*: 389-411.

10.   TAYLOR, B.K. and VAN DEN ENDE, B. (1969) The nitrogen nutrition of the peach tree.  IV: Storage and mobilization of nitrogen in mature trees. *Aust. J. Agric. Res. 20*: 869-881.

# GENERAL DISCUSSION

*An evaluation of the biochemical approach to determine
the conditional status of plants*

F.C. STEWARD (Session Leader): We have had so wide a
range of biochemical observations on the biochemical
consequences of nutritional levels and imbalances that,
as Chairman, it is difficult to select any one topic as
the salient one for discussion. However there has been
heavy interest in the consequences for nitrogen metabolism of the supply of the inorganic nutrients. Reference has been made by several speakers to the now long
established use of chromatography to detect the range
of soluble, or non-protein nitrogen compounds, and
later of gel electrophoresis to detect the range of
more soluble proteins and enzymes, as affected by nutrition. However, even here the hope that a given compound, or biochemical symptom would be a highly nutritional specific for different plants, or the same
one under all conditions, is not to be expected.
Ideally and under optimal balanced conditions the use
of external nitrogen keeps pace with the use of external
carbon by photosynthesis with the maximum conversion of
N to protein.

In fact it is when the pace of growth slows down,
as in storage organs, or the requirements for growth
are dislocated, that the accumulations of chromoto-

251

graphically detectable compounds appears.  We recognized this in the first application of chromatography
in the late 40's in my laboratory.  Thus while we recognize the value of the scanning of the nitrogen compounds
in this way we don't normally ascribe too much diagnostic significance to any one of them e.g. arginine or
glutamine or asparagine as such.  These nitrogen rich
compounds may be caused to accumulate in different ways,
as e.g. by changes in the environment and due to devel-
opmental stimuli.  The point is however that such surveys enable one to scan a wide range of biochemical constituents, quarry clues to alterations in metabolism
which occur in that sensitive area, where carbohydrate
metabolism (in the use of reduction products of carbon
dioxide) and nitrogen metabolism (in the use of reduction products of nitrogen) come together.  But so
long as the other general requirements for growth are
adequate and balanced, these excessive accumulations,
or preponderant emphases on *one* nitrogenous constituent
do not occur.  But they can occur, under optimal nutritional conditions, when for other reasons growth is
slowed down.  When growth is resumed, as in the case of
storage organs, these compounds tend to disappear.  So
we should look to nutrition as but one of the parameters
that will determine how plant cells utilize their gene-
determined biochemical propensities.  Against this back-
ground we have had, and can discuss the following subjects:

a) Observations in several papers on the response
i.e. appearance and disappearance, of individual enzymes
(or their isozymes) in response to supply or removal of
specific elements.  This idea runs through papers by
Bar—Akiva, Bielesky, and at the organelle level by
Possingham.

b) We have had reference to the role of potassium
as an activator of enzyme systems in *in vitro* systems
in general (Wilson), or in the activation (by deficiency or excess) of certain biochemical pathways lead-

ing to amine formation (Hoffman and Samish) which are
dramatically suppressed during optimal growth.

c) We had reference to the special role of potassium in the reciprocal circulation within the plant body
of carbon (via organic acids) from shoot to root and
nitrogen from root to shoot as in the interesting paper
of Lips et al.

d) We have had reference to what we might call the
over-all economy of nitrate utilisation during plant nutrition, as in the papers of Coïc, Sorensen and Taylor.

Thus there are many points for discussion, but we
should be prepared for the greater difficulty in explanation than observation. For plants integrate in their
biochemical expression the effects of many different
gene-regulated consequences of nutrition and environment, as these are to be seen in the nitrogenous compounds (protein and non-protein) which appear and the
reactions they mediate.

DE WIT: There are studies on enzyme level beyond which
we are not possibly interested anymore in the behavior
of plants per se, biochemists or breeders might be interested in that. I am really interested in what goes
on in the initial phases of fertilizer shortage and
where we have slight fertilizer excess. It is these
excesses, where the biochemical reactions are of importance, not when we go to indefinite levels. I don't
really think it is interesting the biochemist but I
think it is particularly uninteresting from the stand-
point of the plant nutritionist.

STEWARD: Professor de Wit emphasized that recovery from
early stages of deficiency is perhaps more interesting
than studying the excessive and irreversible accumulation, because this was not a point which you made with
your inductive enzymes or the isozymes.

BAR-AKIVA: With annual crops, leaf analysis is frequently a post-mortem story indeed, but this is the same
with the conventional leaf analysis and with the biochemical approach. I think there is more chance to
detect the nutrient requirement by using enzyme activity, than with the conventional leaf analysis, - even
with an annual plant.

    With fruit trees, we are not interested only in the
deficient level. As the previous speaker has said, the
growers and farmers can deal with deficiencies themselves. We frequently can't deal with that part of the
standard curve which is above the critical level. Still,
you may sometimes get into trouble even with a slight
excess of fertilizer. Presently we are conducting an
experiment on nitrogen fertilization with Valencia trees
with 3 levels of N. The two higher levels did not bring
any increase in yield, nor any visible deterioration of
the fruit. But, on checking the shelf life of the fruit,
a deterioration of the fruit was observed at the higher
N level, and this is not pocket money. The problem of
the conventional leaf analysis is that it does not mean
too much in the intermediary range of the standard curve,
because actually we have a critical value and not a
range.

    What we are trying to obtain by means of the enzyme
activity studies is the improvement or refinement of
determinations in the intermediary range, where ordinary
leaf analysis at the present time can't do very well.

SAMISH: May I recall to you the curve which you just
saw in connection with the accumulation of putrescine,
which represents a very big advance in that even very
small tendencies towards both K deficiencies and K
excess can be found with the help of the determination
of putrescine when you really have no indication or
symptom whatsoever on the tree itself.

BIELESKI: The most sensitive way of detecting deficiency

we have found with our model system is the effect on
growth rate.  We can detect the effect of phosphorus
removal with spirodela in six hours by a statistically
significant decrease of the growth rate.

At this time there is not as yet a significantly
measurable decrease in the content of phosphate (which
is your leaf analysis) and there has not been a rise in
the phosphatase level, which is also characteristic.
You have to go on about 2 or 3 days to get a really significant decrease in organic phosphate, and to about 5
days to get a really significant increase in phospha-
tases.  I think this is due to the fact that we have
pools of nutrients and that the phosphate which is in
the tissue is not all directly available for growth.
Prof. Steward made these same points about amino acids.
The immediate effect can be just reflecting what is
going on in the metabolic pool and not what is going on
in the whole plant.

## Nitrate utilization

SØRENSEN:  The nitrate level was affected  by the substrates, and is increased by their concentration unto a
certain point, but not further.  We have to think about
the farmers and the production of other things.  It is
very important to speculate about how can we progress
further and how we can bring about a larger increase of
the assimilation of nitrate in the plant.

We have also had to do some investigations concerning individual variations between plants.  We have found
very great variations in the nitrate concentration between plants, and also when comparing nitrogen fractions.
Perhaps it will be the geneticist who should go further
on, but it's a very important point that was raised here.

BOWEN:  In connection with nitrate reductase, some of
you were probably aware of the review by Hageman in 1967

on a genetic basis for physiological selection of
plants.  He points out that in maize there exists considerable variation in nitrate reductase between different lines.  So this provides one possibility, for
example, of increasing nitrate reductase.

JUNCK:  When you have plants with a shortage of molybdenum, then it's well known that they have very high
concentrations of nitrate in the leaves.  These plants
have not only nitrate but also potassium in very high
concentrations.  And when we add molybdenum, we get a
decrease in nitrate, because of reduction, and also we
get a strong decrease in potassium concentration.  This
shows that the transport, retransport, or recirculation
of potassium is only possible when nitrate reduction is
normally functioning.

STEWARD:  There isn't much point in stepping up nitrate
reduction unless you can furnish the rest of the metabolism which puts it into protein.  The more nearly you
get things balanced, the less soluble nitrogen there is
there.  And so you must look at something else, other
parts of the system.  I suppose if trace elements, major
elements, photosynthesis and all the rest of it was working at the optimum balance, then one wonders whether you
need a lot of these soluble nitrogen compounds floating
about.  I don't think you would.

# EVALUATION OF
# THE NUTRITIONAL POTENTIAL OF THE SOIL

Session Leader:  GLYNN BOWEN

# A KINETIC APPROACH TO THE EVALUATION OF THE SOIL NUTRIENT POTENTIAL[*]

Stanley S. Barber

*Dept. of Agronomy, Purdue University, Lafayette, Indiana, U.S.A.*

## ABSTRACT

The chemical potential proposed by Schofield has been investigated by many people in recent years as a method of determining soil nutrient availability. The labeling of nutrient parameters as intensity and capacity has natural appeal. However, do these values actually measure intensity or availability of a nutrient and capacity or ability to maintain a supply to the root? I question this.

There is no question that the chemical potential measurements characterize an equilibrium parameter of the soil. The ability to maintain this in a stirred equilibrium system is also an interesting soil para-

*Journal Paper Number 3928 of the Purdue Agricultural Experiment Station.

Stanley S. BARBER. Prof. Agron., Purdue Univ., Lafayette (Indiana) since 1958. b. 1921 Wolesley (Canada); M.Sc., Univ. Saskatchewan; 1949 Ph.D., Univ. Missouri (U.S.A.); 1949 Asst. Prof. and 1952 Assoc. Prof. Agron., Purdue Univ.

meter. However, plant roots growing in soil are not growing in a stirred equilibrium system. Rates of movement of ions are involved. Nutrients usually must move to the root. These are kinetic mechanisms. The ratio of K/Ca at the root surface is usually very different from that in the equilibrium soil the root is growing in. We have found depletion of K about the root and diffusion governing the rate of supply. Calcium frequently accumulates at the root because mass-flow supplies Ca at a faster rate than the root absorbs it. We have measured increases in soluble Ca at the root surface of 15 times those in the original soil.

In order to measure nutrient availability accurately we must measure the parameters that are important in governing the supply. These will be different for each nutrient. The plant root changes the chemical nature of the rhizosphere soil. It may change the pH by one pH unit or more. It may create either a hydrogen or a bicarbonate gradient out from its surface.

Nutrient absorption from soils is a dynamic process. The parameters that govern these rates will be discussed, as well as the interactions between nutrients caused by the supply mechanisms.

––––––––––

The development of a procedure for characterizing the availability of nutrients in the soil that is based on sound fundamentals is very desirable and it has been the goal of a number of investigations. The procedure should be independent of soil type and apply equally well to all soil-plant root situations. A basic understanding of the processes operating should make it possible to more nearly achieve this goal. Research achievements in recent years have added greatly to this understanding.

In this paper, I do not attempt to review the volu-
minous number of research reports in this field and indicate their contribution.   Instead, I will discuss some
of the approaches to the measurement of the soil nutrient potential that are found frequently in the recent
literature and indicate the relation between them and
the mechanisms I believe to be important in determining
nutrient availability.   I have kept the number of
literature references to a minimum, using only examples
rather than listing all the references.   I hope to
stimulate questions rather than give complete answers.
In addition, I will indicate some of the complicating
factors affecting the measurements of soil nutrient potential.

First, a few definitions are in order.   Soil nutrient potential is a measure of a soil property or
soil properties that define(s) the rate at which the
soil is capable of supplying nutrients to the plant
root.   Soil nutrient availability is the mean rate of
supply of a nutrient to the plant root growing in soil.
These definitions are, I believe, rather similar to the
way in which these terms have been used by a number of
investigators.

Since soil nutrient availability is a rate of supply of a nutrient to the root, it involves both the
initial level of the nutrient at the root surface and
the capacity of the soil to maintain this or some other
level.   It is not surprising that we have parameters
that have been labeled intensity and capacity since
these are important measurements.   However, I question
whether many of the measurements labeled intensity and
capacity actually measure the intensity and capacity
parameters that control soil nutrient availability.


THE INITIAL LEVEL IN SOIL SOLUTION


The chemical potential of P in the soil was

suggested by Schofield (1955) as a measure of P availability and at about the same time Woodruff (1955) proposed the free energy of exchange between K and Ca as a
measure of K availability. Since then, the expressions
$\frac{1}{2}pCa + pH_2PO_4$ and $2.303RT(pK - \frac{1}{2}pCa)$ have frequently
been used as measures of the intensity parameter. In
both expressions, we have the logarithm of the activity
of the ion in question related to the square root of
the Ca activity. Some have used the ratio $K/(Ca)^{\frac{1}{2}}$ or
$K/(Ca + Mg)^{\frac{1}{2}}$ rather than the logarithm of the expression
relating it to chemical potential. Magnesium availability has also been related to $Mg/(Ca + Mg)^{\frac{1}{2}}$ (Salmon,
1964).

In soil systems, the ratio $K/(Ca + Mg)^{\frac{1}{2}}$ is an important parameter to characterize the soil. Since the
ratio remains constant throughout a soil-water system
at equilibrium and does not vary appreciably with dilution, it is a convenient relationship to characterize
the exchangeable K, Ca and Mg. However, what does it
have to do with K availability to the plant root?
Within limits, altering the level of Ca + Mg in solution while holding the level of K in solution constant
has very little effect on the rate of K absorption by
the root, hence the ratio itself has little meaning.
However, the ratio measurement on soil is usually made
after equilibrating with $0.01M$ $CaCl_2$ so that the level
of Ca + Mg is approximately constant for comparisons
between soils. The value causing variation of the
ratio is the level of K. By using a number of solutions containing increasing amounts of K in the $0.01M$
$CaCl_2$ we can, after equilibration with the soil, find
the level of K in solution initially occuring in the
soil (the level where K is neither adsorbed or released
by the soil when $0.01M$ $CaCl_2$ plus K is equilibrated
with the soil).

The rate of absorption of K by the plant root depends primarily on the level of K offered to the root
by the soil. The activity ratio will correlate with

this if the Ca + Mg level remains constant.  Why not
just use the K concentration or activity as the variable instead of the ratio?  We may want to measure the
K in solution by using the same procedures.  There
seems little justification for using ratios or loga-
rithms of ratios.  White (1968) does use P activity,
Ozanne and Shaw (1967) use equilibrium P concentration
in the 0.01M $CaCl_2$ equilibrium solution.  Since flowing
nutrient culture experiments show a close relation between concentration of the nutrient in the solution and
rate of absorption by the root, it would appear that
the concentration or activity of the nutrient at the
root surface would be the appropriate intensity value
to use.


THE MAINTENANCE OF A NUTRIENT LEVEL AT THE ROOT


     While the initial activity of the nutrient at the
root surface may describe the initial rate of absorption by the root, the capacity of the soil to maintain
this or some other level is also important since uptake
proceeds over a period of days or weeks.  There are two
approaches to the measurement of the capacity of the
soil to supply nutrients, one is the capacity-intensity
(sometimes referred to as QI) concept in equilibrium
soil systems; the other is the kinetic approach evaluating nutrient supply mechanisms.  There may be a wide
divergence between the two in the rate of supply to the
root surface that they indicate.

     In the equilibrium system, we can measure the re-
lation between the change of the level of an adsorbed
nutrient on the soil with the change in level of this
nutrient in solution.  In many of these plots, the
activity ratio has been plotted versus the amount of
ion adsorbed or desorbed.  I believe the actual activity
of the nutrient is preferable.  This is an equilibrium
system measurement and when applied to roots growing in

soil it would be appropriate if the roots were growing in a rapidly stirred soil system. However, such is not the case; we do not have equilibrium throughout the soil-root system, but non-equilibrium conditions where nutrients flow toward a sink created by adsorption by the root. After the root has grown into the soil, the movement of nutrients to the root is by mass-flow and diffusion. The movement occurs because the root adsorbs both water and nutrients.

Plant roots absorb water which causes a flow of water to the root. Since the water in soil contains nutrients, this flowing water transports nutrients to the root by mass-flow. If the supply by mass-flow is less than the rate of uptake, the concentration at the root surface is decreased and diffusion toward the root also occurs. If mass-flow supplies more than the root absorbs then the ions accumulate at the root and back diffusion occurs. The rate of uptake by the root will depend on the level at the root which in turn is determined by the balance between uptake rate and the rate of supply by mass-flow and diffusion. Mathematical relations describing the flow to the root as a cylindrical sink have been developed by several investigators (Nye and Speers, 1964; Passioura, 1963). To simplify, I will discuss mass-flow and diffusion as separate supply mechanisms although they do interact.

Mass-flow moves nutrients to the root at the concentration in the flowing soil solution. This is a value that is closely related to the values obtained when we measure the concentrations of P and K in the activity ratio concept. If the root absorbs the nutrient at the same rate as the supply by mass-flow then this parameter of intensity also describes availability if we assume a nutrient that is highly buffered by the soil so that a constant ionic level in solution is maintained. If mass-flow moves larger amounts than is absorbed then the ion accumulates and the uptake rate corresponds to the higher level resulting from accumu-

lation.  If it is less, the concentration at the root
surface drops and diffusion toward the root also occurs.
The rate of uptake would be less than that predicted
from the equilibrium solution concentration because of
the reduction in concentration at the root.  The level
of soluble anions in the soil affect the rate of supply
of cations by mass-flow.  Nitrate is frequently a dominant anion in well fertilized soils.  If organic matter
mineralization maintains the level of nitrate in the
soil solution, the rate of supply of cations by mass-
flow in a soil highly buffered with these cations should
remain relatively constant.  Where the buffering capacity is low or the anion content of the solution
drops, supply by mass-flow will decrease.

     Diffusive flow of nutrients to the root depends
upon the rate of diffusion and concentration gradient.
Diffusion usually supplies most of the P since mass-
flow seldom supplies more than a few percent of the
requirement.  The rate of supply by diffusion has been
approached in two ways.  One is to measure the labile P
adsorbed in the soil to get the concentration gradient
and use the apparent rate of diffusion for the diffusion
coefficient.  The apparent diffusion coefficient in soil
is less than that in water principally because of three
factors; the volumetric moisture content, the tortuosity
and the buffering capacity of the soil.  The latter is
usually by far the most important since the ratio of
soluble P to adsorbed P in most soils is of the order of
1-1,000 to 1-10,000.  The second approach is to use the
concentration gradient of the P in solution and to use
the rate of diffusion of the nutrient in water for the
diffusion coefficient.  The resulting rate of diffusion
in the soil solution is then affected mainly by the
volumetric moisture and tortuosity and not the buffering
capacity.  In this approach, the gradient is smaller by
a factor equal to the buffering capacity and the diffusion coefficient is greater by the same factor.  The
supply by diffusion can be, as a close approximation,
described by either.  The concentration gradient and

fusion coefficient are then used in the appropriate
equation describing diffusive flow to a cylindrical
sink.  Since the concentration in solution is used in
the second approach, the concentration gradient for diffusion is related to the intensity measurement of P
where the concentration  or activity in solution is
measured.  Thus one of the parameters for calculation
by diffusion is the same as that in the calculation of
P intensity.

The use of solution P works well as long as there
is no overlap between soil diffusion diameters (the diameter of soil about a root from which diffusion occurs
to an appreciable extent).  So we might suggest that
where soil diffusion diameters are small relative to
inter-root distances, the intensity measurement of the
concentration of a nutrient, such as P, in solution may
be a good measure of the supply of P to the root by diffusion.

The soil diffusion diameter is inversely related to
the buffering capacity, so for P the value may be only
of the order of one mm since P usually has a high buffering capacity.  For K, where the buffering capacity
term is much smaller, of the order of 50-100, the soil
diffusion diameter is larger and overlap between adjacent roots is likely to occur.  When overlap is predominant, the buffering capacity or more likely, the
labile ion content of the soil, will become more important in determining the rate of supply to the root
by diffusion.  With the smaller buffering capacity, the
root usually adsorbs a much larger portion of the total
labile quantity present.

In addition to mass-flow and diffusion, nutrients
may reach the root by root interception.  As the root
grows through the soil it occupies space; usually about
one percent of the total that was initially occupied by
soil.  The root absorbs or displaces the nutrients in
this volume.  This mechanism is more important where the

labile concentration in the soil is highly relative to the amount absorbed (for example, Ca and Mg on soils with large amounts of exchangeable Ca and Mg). The amount supplied by root interception is related to the total labile content of the soil.

Let's see how these ideas relate to observation. Ozanne and Shaw (1967) measured the P content in solution in equilibrium with 0.01M $CaCl_2$. This is presumably closely related to the P level in solution in the soil at equilibrium. They observed a close correlation between the relative yield of pasture and this equilibrium P concentration. When they determined the amount of P sorbed by the soil at an equilibrium concentration of 0.30 ppm and plotted this against the P required to get 95% of maximum yield they got an r of 0.91 for the combined data from 21 sites in 1963 and 18 different sites in 1964. While this may indicate the significance of the level in solution is an appropriate parameter, we also must recognize that most soil testing procedures for P in use today remove a portion or all of the labile P in the soil (some also remove non-labile P). The fact that they give close correlation with P availability within local areas indicates either that the labile or adsorbed P is important or that there is a close correlation between solution P and labile P. Recent information (Nagarajah et al., 1968) indicates a hysteresis between adsorption and desorption which may affect availability so that the desorbing ions present in solution may also influence availability of P.

Investigations of K availability in the soil have often shown a close correlation to exist between exchangeable K and K uptake by the plant. One study of this type made on 38 surface (0-6 inches) and subsurface (18-24 inches) soils from 6 states in the midwestern region of the United States gave an r of 0.96 between the exchangeable K and the K adsorbed by millet. A study involving 52 soils, conducted in the preceding

year gave an r of 0.88.  These results indicated that
the total adsorbed or exchangeable K was a suitable
measure.  Others have not obtained as high correlation
of plant uptake K with exchangeable K.  Arnold (1962)
obtained a correlation of r = 0.35 when he used the
range 10-20 mg exchangeable K per 100 g.  By using the
free energy of exchange between K and Ca he obtained an
r of 0.64.  He did however get an r of 0.88 when he
correlated K uptake by ryegrass with exchangeable K of
all the soils rather than those in a restricted range.

In our research at Purdue University, we have shown
how the rate of transpiration affects those nutrients
supplied by mass-flow but not those supplied by diffusion.  A three-fold increase in transpiration rate
doubled Ca absorption but had little effect on K absorption.  Calculations of the amount supplied by mass-flow
indicate that supply of Ca by mass-flow exceed uptake
whereas supply of K by mass-flow was less than 10 percent of uptake.

We were also able to show a close relation between
the supply of $^{86}$Rb by diffusion and the rate of absorption of $^{86}$Rb by corn.  **The** rate of supply was varied by
changing the moisture content, the clay content and the
level of Rb in the soil.

The supply of nutrients to the plant root is undoubtedly complex.  The ideas I wish to present are
that rate processes affect the rate of supply of nutrients to the root and hence their availability.
There is a rate process determining supply by diffusion and an independent rate process determining the
rate of supply by mass-flow.  The buffering capacity
of the soil affects both of these, but not to the same
degree.

The appropriate procedure for characterizing the
soil nutrient potential will undoubtedly vary with the
nutrient.  The relative significance of mass-flow and

diffusion as supply mechanisms will modify the procedures to be used for a particular nutrient.  When we
consider the two nutrients, Ca and K, we usually find
Ca supplied predominantly by mass-flow and K by diffusion.  We need to treat them separately.  These two
nutrients illustrate the difficulties of using a ratio
of $K/(Ca)^{\frac{1}{2}}$ because the ratio at the root surface may
be different by a factor of 10 or more from the equilibrium ratio in the soil.  The K will be depleted at
the root because the rate of supply by mass-flow is
much less than uptake rate whereas Ca may accumulate.

    We could use equations describing supply and
measure all the parameters needed.  However, I believe
we can simplify this by measuring only those parameters
that dominate the rate of supply.


PLANT ROOTS ALTER THE RHIZOSPHERE


    Plant roots are not merely sinks for the movement
of water and nutrients, they also modify the chemistry
and biology of the rhizosphere soil.  Because of this
they may influence the level of nutrients released from
the soil and the distribution between solution and adsorbed phases.  These effects may be similar for the
same variety of a species but can vary between species.

    Some of the effects that plant roots have on
their rhizosphere are as follows: (1) They increase
soil density because they force their way through the
soil.  Lund *et al*. (1965) have shown this effect in
photographs of soil-root thin sections.  (2) They may
change the salt concentration because of greater flow
to the root by mass-flow than absorption.  Riley and
Barber (1970) have shown 7-fold increased in soluble
anion contents.  (3) They change the soil pH.  Riley
and Barber (1969) and Ozanne and Barber (1970) have
shown increases and decreases of magnitude of one pH

unit each way. This results from an imbalance between cation and anion uptake. When cation uptake exceeds anion uptake $H^+$ is released by the root and the pH drops. When the reverse occurs, $HCO_3^-$ is given off and the pH increases. (4) Changes may occur in the relative levels of the cations and anions both because of differential rates of absorption as compared to supply and because of release of $H^+$ or $HCO_3^-$.

The modification of the rhizosphere soil by the root is likely to result in changes in the rate of supply of nutrients to the root. These modifications may cause changes in supply both by mass-flow and diffusion. The solubility of many ions in soil solution is pH dependent. The root is also known to change the level of microbial activity in its vicinity and this may change nutrient availability. I believe that the chemical effect is greater than the microbiological effect.

SUMMARY

In summary, the plant root acts both as a sink for nutrients absorbed by the plant and as a modifier of the physical-chemical and biological properties of the soil adjacent to the roots. Nutrients flow to this sink by mass-flow and diffusion. The principal supply mechanism will determine to a large degree the parameter which should be used to measure the soil nutrient potential. Where mass-flow is the main supply, the level in the flowing solution would be a realistic parameter. Where diffusion is the principal supply mechanism, the parameter to measure will depend upon the root density and the apparent diffusion rate. Where soil diffusion diameters do not overlap appreciably, then the level of the ion in the solution may be an appropriate parameter. Where overlap is significant then the labile level of the nutrient in the soil may be more important.

Roots do modify the rhizosphere soil. Differences may occur between species and differences are also caused by variation in the relative concentrations of the nutrients in the soil.

REFERENCES

1.    ARNOLD, P.W. (1962) The potassium status of some
      English soils considered as a problem of energy
      relationships. *Proc. Fert. Soc. 72*: 25-54.

2.    BARBER, S.A. *et al.* (1961) North Central Regional
      Potassium Studies. II: Greenhouse experiments
      with millet. *Research Bulletin. 717*, Purdue
      University Agricultural Experiment Station,
      Lafayette, Indiana.

3.    BARROW, N.J. (1967) Relationship between uptake of
      phosphorus by plants and the phosphorus poten-
      tial and buffering capacity of the soil - an at-
      tempt to test Schofield's hypothesis. *Soil Sci.
      104*: 99-106.

4.    LUND, Z.F. and BEALS, H.O. (1965) A technique for
      making thin sections of soil with roots in
      place. *Soil Sci. Soc. Amer. Proc. 29*: 633-635.

5.    NAGARAJAH, S., POSNER, A.M. and QUIRK, J.P. (1968)
      Desorption of phosphate from koolinite by
      citrate and bicarbonate. *Soil Sci. Soc. Amer.
      Proc. 32*: 507-510.

6.    NYE, P.H. and SPIERS, J.A. (1964) Simultaneous
      diffusion and mass-flow to plant roots. *Trans.
      8th Congr. Int. Soil Sci. Soc. Bucharest. 3*:
      535-542.

7.   OZANNE, P.G. and BARBER, S.A. (1970) The influence
     of nitrogen source on the degree to which cape-
     wood, ryegrass, lupin and subterranean clover
     modify their rhizosphere soil. *Aust. J. Agric.
     Sci.* (In press.)

8.   OZANNE, P.G. and SHAW, T.C. (1967) Phosphate sorp-
     tion by soils as a measure of the phosphate re-
     quirement for pasture growth. *Aust. J. of
     Agric. Res. 18*: 601-612.

9.   PASSIOURA, J.B. (1963) A mathematical model for
     the uptake of ions from the soil solution.
     *Pl. Soil. 18*: 225-238.

10.  RILEY, D. and BARBER, S.A. (1969) Bicarbonate ac-
     cumulation and pH changes at the soybean
     (*Glycine max.* (L.) *Merr.*) root-soil interface.
     *Soil Sci. Soc. Amer. Proc. 33.*  905-908.

11.  RILEY, D. and BARBER, S.A. (1970) Salt accumulation
     at the soybean (*Glycine max.* (L.) *Merr.*) root-
     soil interface. *Soil Sci. Soc. Amer. Proc. 34.*
     154-155.

12.  SALMON, R.C. (1964) Cation activity ratios in
     equilibrium soil solutions and the availability
     of magnesium. *Soil. Sci. 98*: 213-221.

13.  SCHOFIELD, R.K. (1955) Can a precise meaning be
     given to "available" soil phosphorus? *Soils
     and Fertilizers 18*: 373-375.

14.  WHITE, R.E. (1968) Buffering capacity of soil on
     uptake of phosphorus by plants. *9th Int. Cong.
     Soil Sci., Adelaide II*: 787-794.

15.  WOODRUFF, C.M. (1955) The energies of replacement
     of calcium by potassium in soils. *Soil Sci.
     Soc. Amer. Proc. 19*: 167-171.

*Questions to Prof. Barber.*

BIELESKY: Could Prof. Barber give us an indication of the techniques he has used in obtaining autoradiographs of the rhizosphere?

BARBER: Plants are grown in a box 2-3 mm wide and with one side made of thin polyethylene film. The soil in the box is uniformly labelled and X-ray film is placed against the polyethylene film (in the dark room) after varying periods. This shows how plant roots growing next to the polyethylene film absorb the radioactivity and alter the distribution of radioactivity in soil around them. The autoradiograph measures only a very thin section of the soil and roots close to the surface, the thinness, of course, depending on the relative strength of the β particle of the ion being measured.

In other systems we are more directly measuring ions in the rhizosphere soil by carefully picking the soil off the roots with tweezers.

NIELSEN: Do you find a higher percent of phosphorus in the tissue when you use ammonia than when you use nitrate?

BARBER: It appears that the higher phosphorus uptake with ammonia is because of the effect of ammonia on the balance of a cation to anion uptake, and the resulting reduction in pH of the soil around the root. When you reduce the pH in some soils you increase the phosphorus in solution and increase P availability.

When the plant is grown in solution, the P difference between ammonia and nitrate supply is not found.

# COMPUTER MODELING OF NUTRIENT MOVEMENT IN SOILS

Maurice H. Frere

*U.S. Soils Laboratory, Agricultural Research Service,
U.S.D.A., Beltsville, Maryland*

Cornelius T. de Wit

*Agricultural University, Wageningen, The Netherlands*

ABSTRACT

The movement of nutrients through the soil to the
plant root is an important consideration in the nutrition of plants.   Ion species differ in their mobility, concentration, adsorption properties and uptake
rates, yet the soil-plant system remains electrically
neutral.  This illustrates the complexity of the interactions which often occur so close to the root surface that measurements of them are extremely difficult.

Mathematical analysis can be used profitably as
a tool to predict and evaluate the relative importance of the various factors.  The usefulness of mathematical analysis is usually limited by the extent
to which the mathematical model corresponds to the
real system.  Modern digital computers can be used
not only to obtain numerical solutions of differential
equations, but also to simulate, through numerical
calculation, the behavior of the system under a

Cornelius T. DE WIT.  Extraord. Prof. Theoret. Prod. Ecol., Agric. Univ., Wageningen (Netherl.) since 1968.
b. 1924 Brummen (Netherl.); 1950 M.Sc. and 1953 Ph.D., Agric. Univ., Wageningen; 1952 Min. Nat.
Planning (Burma); 1956 Researcher, Inst. Biol. Chem. Res. Field Crops & Herbage, Wageningen.

variety of condition.  This technique usually requires
less restrictive assumptions and thus permits the model
to be more complex and realistic.


THEORY


     The simplest simulation approach is analogous to
bookkeeping, where each account loses and gains value
in a variety of ways and rates.  The root is visualized
as an account that takes some ions from the adjoining
soil and releases others to maintain electrical
neutrality.  The soil is divided into a number of
accounts that correspond to concentric cylinders of
increasing size surrounding the root, Fig. 1.  The ions
move into each cylinder from the adjacent cylinders
according to the flow of water and the differences in
the concentrations that exist.  After each small increment of time in which movement occurs, each account
is updated, not only for the total amount of each ion
that is present but also for the distribution of each
between the adsorbed and solution phases.

     The equation for the movement including terms for
mass flow and electro-chemical force, is calculated from
the concentrations and concentration gradients of all
ion species.  Mass action equations for cation exchange
between the solution and the soil are used to equi-
librate each cylinder between periods of movement.

     The simulation calculations, whose arrangement is
shown in Fig. 2, start with the first or innermost
cylinder.  The amount of each ion taken up or liberated
by the root is calculated.  Then the amount moved
between the first and second cylinders is calculated.
Finally the new concentration is calculated by adding
to the old concentration the net change in the amount
of the ion divided by the volume of the cylinder.
For the second cylinder the amount moved to or from the

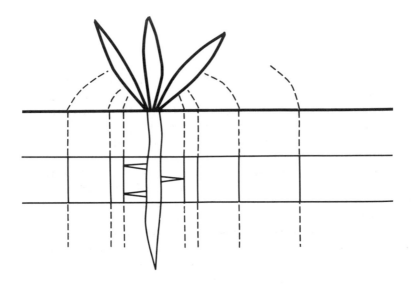

FIG. 1.  A diagram of the soil around a unit length of
         root divided into concentric cylinders.

first cylinder has already been calculated and only the
amount to or from the third cylinder needs to be calculated.  In this way the calculations progress from
cylinder to cylinder out to the last one.  Since the
rates are calculated and then the concentrations are
changed, this method of integration corresponds with the
simple point-slope method of Euler.  In this model it
is assumed that the concentration beyond the last
cylinder never changes.  An alternate assumption would
be that the last cylinder is midway between two roots

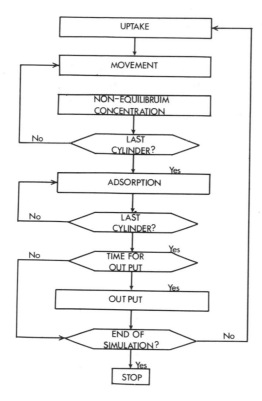

FIG. 2.  The flow diagram of the calculations for
         simulating nutrient movement through the
         soil to plant roots.

and moves equal amounts of ions in both directions.

        After a new concentration is calculated for each
cylinder, the new equilibrium between the adsorbed and
solution phases is calculated for each cylinder.  Before
returning to the uptake calculations for a new increment
of time, the accumulated time increments are checked to
determine if it is time to print the current status of
the system and/or time to stop the simulation.

RESULTS AND DISCUSSION

The change in the concentrations of the nutrients
in the root hair zone as a function of time is illustrated in Fig. 3 and the change in the concentration
with distance from the root surface after 1-hour is
illustrated in Fig. 4.

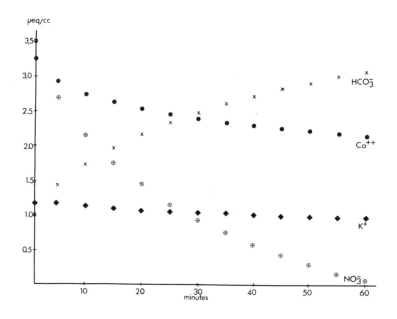

FIG. 3.   The nutrient concentration in the solu-
          tion of the root hair zone as a function
          of time with zero water flow.

The buffering effect of the exchange capacity on the
cation concentrations compared to the rapid depletion
of the nitrate concentration is clearly illustrated in
these figures.  This was suggested some time ago by Bray
(1954) but the actual time period involved, about an

hour, and the resulting steep gradients can only be obtained from actual numerical examples.

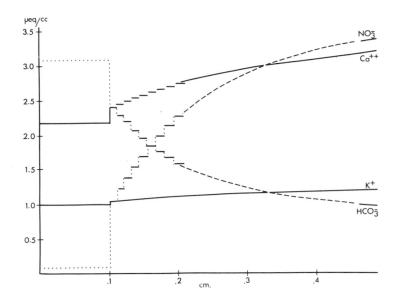

FIG. 4.    The nutrient concentration as a function
           of the distance from the root surface
           after 1-hour of simulated time with zero
           water flow.

The rates of uptake used are moderate and correspond to each centimeter of root supporting a growth rate of 7 mg of dry matter per day containing 1.9% N, 2% K, and 0.35% Ca.

The relative importance of different factors in the system can be studied by repeating the simulations with different values for those factors.   An example of this is a recent study of the electrical interaction between ions (Frere, 1969).

REFERENCES

1.    BRAY, H.R. (1954)  A nutrient mobility concept of
      soil-plant relationships by cylindrical roots of
      onion and leek.  *Soil Sci. 78*: 9-22.

2.    FRERE, M.H. (1969)  Ionic interaction in diffusion.
      *Soil Sci. Soc. Amer. Proc. 33*: 883-886.

*Questions to Prof. de Wit*

BARBER:  I note that you did not have mass flow operating in your simulation.

DE WIT:  Yes.  In fact, if mass flow is operating,
calcium around the root is much higher.

# A NEW APPROACH TO EVALUATING THE EFFECTIVE ROOT AREA OF FIELD CROPS

Benjamin Bar-Yosef and Uzi Kafkafi

*The Volcani Institute of Agricultural Research, Bet Dagan, Israel*

## ABSTRACT

The movement of ions in the soil towards roots and the evaluation of diffusion processes in nutrient uptake by plants have lately been receiving increased attention. Amongst the parameters needed to solve and interpret the diffusion equation involved, the surface area of the root is the most difficult one to measure. Such measurements based on the direct measurement of the entire root system are carried out successfully under laboratory conditions on small seedlings. However, the estimation of the root surface area of a growing plant in the field with this method is tedious, and the method cannot distinguish between effective absorbing roots and non-active roots.

This work presents an approach to evaluating the effective root area (ERA) of a growing crop under field conditions. It is based on the exclusive characteristics of $Ca^{++}$ in calcareous soils, of which the high buffer capacity for $Ca^{++}$ keeps a high and constant calcium concentration in the soil solution. The flux of calcium to the plant under such soil conditions during brief periods of uptake of seven days is used as a

Uzi KAFKAFI. Head Dept. Soil Chem. & Plant Nutr., Volcani Inst. Agric. Res., Rehovot (Israel) since 1970. b. 1934 Tel Aviv; 1959 M.Sc. and 1963 Ph.D. (Agr.), Heb. Univ. (Rehovot); 1959 onwards Res. Soil Scientist, Volcani Inst.

parameter in the calculation of "Effective Root Area"
according to the equation:

$$ERA_i = \Delta Ca_i / F_{Ca}$$

where $\Delta Ca_i$ is the amount of Ca accumulated in one plant
during any week i of the plant's growth, expressed in
$\mu g Ca/plant/week$ and $F_{Ca}$ is the average experimental value
of $Ca^{++}$ flux to the roots in $\mu g Ca/cm^2/week$. The units
for ERA are $cm^2/plant$.


INTRODUCTION


    The movement of ions in the soil towards roots and
the evaluation of diffusion processes in nutrient uptake
by plants have lately been receiving increased attention
(Olsen and Kemper, 1968; Nye, 1966; Lewis and Quirk,
1967). Among the parameters needed to solve and interpret the diffusion equation involved, the surface area
of the root is the most difficult to measure. Such
measurements have been carried out successfully under
laboratory conditions on small seedlings (Russel and
Sanderson, 1967). However, estimation of the root surface area of growing plants in the field is much more
difficult, and only a few attempts to solve this problem are recorded in the literature (e.g. Lewis and
Quirk, 1967; Weaver, 1926).


MATERIALS


    Corn was chosen as a model plant and was grown un-
der field conditions on several levels of soil fertility
of which, for the sake of clarity, only two (low, no
fertilizer applied; and high, 2000 kg/ha superphosphate,

and 2000 kg/h ammonium sulfate added) are presented.
Each treatment was replicated four times. The soil had
a cation exchange capacity of 50 meq/100 g, 75% of which
was $Ca^{++}$, and 8% $CaCO_3$. The above-ground parts of the
plants were sampled weekly and analyzed for Ca and P.
The soil moisture was kept close to field capacity
throughout.


METHOD OF CALCULATIONS


The method used to calculate the effective root
area (ERA, $cm^2$/plant) is based on the ratio (1)

$$ERA_i = \frac{\Delta Ca_i}{F_{Ca}} \qquad (1)$$

between the total amount of Ca taken up during brief absorbing periods of one week ($\Delta Ca_i$, µg/plant) and exper-

imentally known average values of Ca flux to the roots
($F_{Ca}$, µgCa/$cm^2$/week) obtained under Ca concentration con-

ditions similar to those found in the soil. Only meager
information concerning Ca flux through corn roots was
found in the literature (Sabet and Abdel Salem, 1966;
Loneragan and Snowball, 1969). As the root surface area,
rather than its weight, is needed for the calculation of
ion diffusion towards the root, a value of 120 $cm^2$/g
root fresh weight (according to Russel and Sanderson,
1967) was used in the calculations.

Recalculation of the data of Sabet and Abdel Salem
(1966) at Ca concentration of $0.5 \times 10^{-2}$ eq/1 in flowing
solution gave an average Ca-flux value of 2.0 - 2.4 µg
Ca/$cm^2$ root/day. This value is also in agreement with
the results obtained by recalculation of the data of
Loneragan and Snowball (1969).

In order to compare the results of the reported

nutrient solution experiments with field observations, we have combined Weaver's data with ours. Weaver (1926) has reported root surface area for 5-week-old corn plants. Knowing the total Ca uptake during the same period in our experiment, we calculated the average flux value of Ca to the plant root to be $1.8 - 2.2$ $\mu g/Ca/cm^2/day$.

Since weekly samples were taken in our field experiment, an average value of 15 $\mu g Ca^{++}/cm^2/week$ was chosen to characterize the $Ca^{++}$ flux to the corn root.

The ratio $\dfrac{\Delta Ca_i\,(\mu g Ca/plant/week)}{15\,(\mu g Ca/cm^2/week)} = ERA_i\ (cm^2/plant),$

where $\Delta Ca_i$ is the amount of Ca accumulated in one plant during week i, would yield the effective root area ($ERA_i$ $cm^2/plant$) of the plant in the ith week, i ranging from 1 to 11 weeks during the corn's economic growth period. The amount of Ca stored in the roots was ignored, as Weaver (1926) reported that the root dry matter amounts to only 7-8% of the dry weight of the above-ground parts.

RESULTS AND DISCUSSION

The calculated mean weekly addition of effective root areas ($ERA_i$) of corn, plotted against the plant age, is presented in Fig. 1 for the two levels of fertility. Fig. 1 indicates that during the first 40-50 days the effective root area increased gradually. After this period the increments to the root system gradually became smaller. The sum of the points on the reported curve should yield the total effective root area of the plant ($\Sigma[\Delta ERA]$). This value represents the total area which participated in uptake during the entire growing

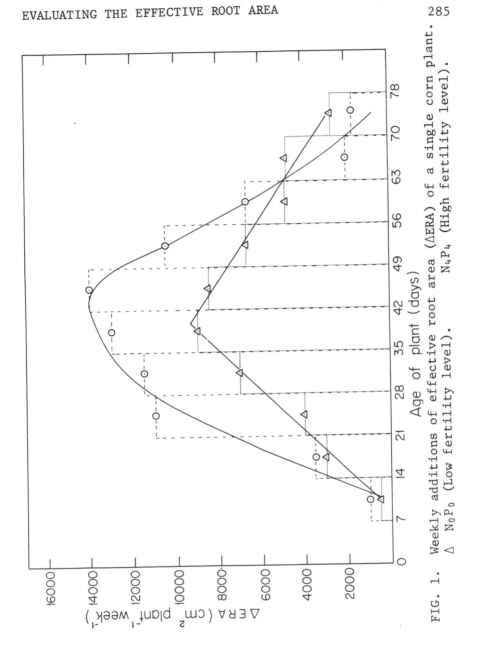

FIG. 1. Weekly additions of effective root area (ΔERA) of a single corn plant.
○ $N_4P_4$ (High fertility level).
△ $N_0P_0$ (Low fertility level).

period.  Using the model for corn root as suggested by
Olsen and Watanabe (1966), we calculated the total
length of the roots of a single corn plant in our field
experiment to be 1200 m and 810 m for the higher and
lower fertility level treatments, respectively.  These
values are of the same order of magnitude as those found
by direct observation of root systems (Weaver, 1926).

The present approach to evaluating the effective
root area (ERA) of a growing crop under field conditions
is similar in concept to that of Lewis and Quirk (1967).
It differs, however, in two main aspects; (a) the $Ca^{++}$
uptake rate, instead of that of phosphate, is used to
estimate the ERA; and (b) an average experimental value
is assigned to $Ca^{++}$ flux to the root during a brief
period of uptake, instead of complicated calculations of
phosphate flux values.

The suggestion that the uptake rate of $Ca^{++}$ can
serve as a better index of ERA is supported by the
following: (a) soil systems generally have a high buffer
capacity for $Ca^{++}$ (Russel, 1961); (b) the uptake of $Ca^{++}$
by the crop is relatively small; and (c) mass flow can
potentially bring to the root surface six times more
$Ca^{++}$ than is found in the plant, which indicates that
the absorption is controlled by the root membrane.  The
$Ca^{++}$ concentration in the soil solution at the root surface may therefore be considered essentially constant.
This cannot be assumed in the case of phosphate (Bouldin,
1961).

Due to these exclusive characteristics of $Ca^{++}$ in
the soil, under conditions where the soil is kept most
of the time close to field capacity, the flux of $Ca^{++}$
through the root membrane can be taken as constant
during a brief absorption period of seven days (Rovira
and Bowen, 1968).  At the least, we assume that the variations in the $Ca^{++}$ flux are not greater than the experimental errors involved in field sampling.

TABLE 1. The Weekly Average Calculated Flux of P to Corn Roots ($\mu gP/cm^2/week$)

| Fertility level | Age of Plants, in weeks | | | | | | | | | | Total average weekly flux |
|---|---|---|---|---|---|---|---|---|---|---|---|
| | 2 | 3 | 4 | 5 | 6 | 7 | 8 | 9 | 10 | 11 | |
| | P flux ($\mu gP/cm^2/week$) | | | | | | | | | | |
| Low | 5.3 | 5.0 | 6.1 | 6.1 | 6.4 | 7.6 | 6.0 | 5.4 | 6.2 | 4.9 | 5.9 |
| High | 6.2 | 9.1 | 7.2 | 8.0 | 8.0 | 8.3 | 9.5 | 9.0 | 7.5 | 6.0 | 8.0 |

The ERA obtained by the Ca flux estimation may be useful in calculating average fluxes of other nutrients for which it is not justified to assume constant concentration. The errors in the flux calculations for any nutrient (X), based on the ERA, are expected to be smaller than the errors introduced to the ERA calculation itself, if the following relation exists:

$$\frac{\% \text{ Ca in top}}{\% \text{ Ca in root}} = \frac{\% \text{ (X) in top}}{\% \text{ (X) in root}}$$

Such a calculation is demonstrated for P in Table 1. The weekly uptake of P per plant ($\Delta P_i$, $\mu gP/plant/week$) was determined. The ratio $\frac{(\Delta P)i}{(\Delta ERA)i}$ was calculated on the assumption that the same root area absorbs P as well as Ca (Russel and Sanderson, 1967). The results yield the average flux of P to the roots ($\mu gP/cm^2/week$) in the ith week. Flux values of P along the growing season are summarized in Table 1.

The results given in Table 1 indicate that the flux values of P tend to remain constant throughout the growth period of the plant, in agreement with the observations of Lewis and Quirk (1967) and of Olsen and Kemper (1968). The total average flux of P at the higher level of fertility was found to be 8.0 $\mu gP/cm^2/week$, as compared to 5.9 at the lower fertility level. The P flux values obtained are in agreement with those reported for corn grown in a greenhouse (Fried and Broeshart, 1967; Hagan and Hopkins, 1955), thereby pointing the way towards a more sound extrapolation of greenhouse experiments to field conditions.

The knowledge of such average flux values can provide availability indices for certain ions, and may help to define boundary conditions in diffusion equations describing ion movement towards roots.

ACKNOWLEDGMENTS

The authors would like to thank Dr. Y. Vaadia and Dr. J. Putter for stimulating discussions and helpful suggestions.

## REFERENCES

1. BARBER, S.A., WALKER, J.M. and VASEY, E.H. (1963) Mechanism of plant nutrients movement from the soil and fertilizer to the plant root. *J. Agric. Ed. Chem.* *11*: 204-7.

2. BOULDIN, D.R. (1961) Mathematical description of diffusion processes in the soil-plant system. *Proc. Soil Sci. Soc. Am.* *25*: 476-80.

3. FRIED, M. and BROESHART, H. (1967) Plant Soil System. Academic Press, New York, p. 64.

4. HAGAN, C.E. and HOPKINS, H.T. (1955) Ionic species in orthophosphate absorption by barley roots. *Pl. Physiol.* *30*: 193-9.

5. LEWIS, D.G. and QUIRK, J.P. (1967) P diffusion in soil and uptake by plant. IV: Computed uptake by model roots as a result of diffusion flow. *Pl. Soil 26*: 454-68.

6. LONERAGAN, J.F. and SNOWBALL, K. (1969) Rate of calcium absorption by plant roots and its relation to growth. *Aust. J. Agric. Res. 20*: 479-90.

7. NYE, P.H. (1966) The effect of nutrient intensity and buffering power of a soil, and the absorbing power, size and root hair of a root, on nutrient absorption by diffusion. *Pl. Soil 25*: 81-105.

8.   OLSEN, S.R. and KEMPER, W.D. (1968)  Movement of
     nutrients to plant roots. *Adv. Agron. 20:*
     91-151.

9.   OLSEN, S.R. and WATANABE, F.S. (1966)  Effective
     volume of soil around plant roots determined
     from P diffusion. *Proc. Soil Sci. Soc. Am. 30:*
     293-302.

10.  ROVIRA, A.D. and BOWEN, G.D. (1968)  Anion uptake
     by plant roots: distribution of anions and ef-
     fect of micro organisms. *9th Int. Conf. Soil
     Sci., Adelaide, Aust. II:* 209-217.

11.  RUSSEL, E.W. (1961)  Soil Condition and Plant Growth.
     9th ed. Longmans Green & Co., London, p. 106.

12.  RUSSEL, R.S. and SANDERSON, J.J. (1967)  Nutrient
     uptake by different parts of plants. *J. Exp.
     Bot. 18:* 491-508.

13.  SABET, S.A. and ABDEL SALEM, M.A. (1966)  Growth
     and ion uptake by maize seedlings in solution
     variable in calcium and flow rate. *Pl. Soil
     24:* 467-74.

14.  WEAVER, J.E. (1926)  Root Development of Field
     Crops. McGraw-Hill, New York.

*Questions to Dr. Kafkafi*

WAISEL:  Would you agree that this method would be valid only under constant water stress, under constant water supply and transpiration?

KAFKAFI:  Yes.  We calculated effective root area for a well grown crop in the field under irrigation.  We kept the soil moist most of the time, and it was never below about 30% of the field capacity.

In addition, I think that in order to get a good evaluation of soil fertility we have to assume that only nutrients are the limiting factor.  If water comes into the picture then, of course, the yield will depend on water, and not all the nutrients which are going into the plant.

BOWEN:  It seems to me that you have a number of assumptions built into the open method for effective root area:  (i) you are calculating the total on entry into the plant by the amount in the top rather than top plus root.  (ii) you are assuming that the effective area for calcium is the same as the effective area for other ions (i.e. all ions are entering the root at the same rates) which may not be so, and (iii) you assume that the percentage of calcium transported to the tops is the same for all ions.  There are data indicating that this is not so.

KAFKAFI:  There are many assumptions here, and we took the best estimates we could from the literature.  The point is that we want to reach, by nutrition, the maximum possible yield produced by nutrients, and if we have enough phosphorus and calcium, the ratio is kept constant in the shoots.

The present work gives us the point of departure for further research, and we can include a correction factor, but the idea is not changed by this.

# RELATIONSHIP BETWEEN ROOT MORPHOLOGY AND NUTRIENT UPTAKE

## Glynn D. Bowen and Albert D. Rovira

*CSIRO, Division of Soils, Glen Osmond, South Australia*

ABSTRACT

The exploitation of plant nutrients from soil and applied fertilizer by roots is dependent upon the morphology of the root system and the ability of different regions of roots to take up and translocate nutrients to the tops.

Although there is considerable information on the gross uptake and translocation of ions by plant roots, there are relatively few studies which specify the activities of various parts of roots.

We have used a technique in which roots are pulse labeled with radioactive phosphate, chloride, sulfate or zinc. The radioactivity along the roots is then measured by passing them through a recording radio chromatogram scanner. This technique has shown that lateral roots of the seminal root system of wheat seedlings, grown in soil for 14 days, account for over 70% of the uptake of sulfate and phosphate. Proportionally more phosphate is translocated from this lateral root zone to the tops than from the apical region of the roots.

Glynn D. BOWEN. Princ. Res. Scientist, CSIRO Div. Soils, Adelaide, (Austral.) b. 1930 Queensland (Austral.);
1953 B.Sc. and 1961 M.Sc., Queensland Univ.; 1947 Dept. Prim. Indus., Queensland; 1958 Res. Scientist,
CSIRO Div. Soils.

Soil temperature had a dramatic effect on root growth and morphology. Wheat roots grown at 10°C averaged 16 cm in length with 3-5 laterals compared with roots grown at 20°C which were 27-38 cm long with 57-82 lateral roots per main seminal root. Translocation of $^{32}$P- phosphate to tops following pulse labelling was four times faster at 20°C than at 10°C.

The importance of lateral roots in the uptake and translocation of phosphate and the effect of soil temperature on root length and morphology raises the question of the possibility of breeding plants with specific types of root systems. If plant breeders can produce plants which have abundant lateral root production at realistic soil temperatures we should thereby obtain a much more efficient use of natural nutrients in the soil and also of applied fertilizer.

## INTRODUCTION

The exploitation of plant nutrients from soil and applied fertilizer by roots is dependent upon the morphology of the root system and the ability of different regions of roots to take up and translocate nutrients to the tops.

Although there is considerable information on the gross uptake and translocation of ions by plant roots, there are relatively few studies which specify the activities of various parts of roots.

In our technique roots are pulse labelled with radioactive nutrients and the radioactivity along the roots measured by passing them through a recording radio-chromatogram scanner (Bowen and Rovira 1967, 1969; Rovira and Bowen 1968 a, b). This technique, with appropriate modifications, has enabled us to study

the influences of environment and plant species on the
patterns of uptake, loss, translocation and biochemical
incorporation along roots (Bowen 1968, 1969, 1970;
Rovira 1969). The results presented here deal mainly
with wheat (*Triticum aestivum* L. var. Gabo) but essentially similar results occur with seedlings of *Pinus
radiata.*


ROLE OF LATERAL ROOTS IN NUTRIENT UPTAKE


Wheat roots grown in soil at 20°C for two weeks
were carefully removed, washed free of adhering soil
and labelled for 15 minutes with radioactive phosphate,
sulfate or chloride as described in Rovira and Bowen
(1968 a). The roots were then washed for 5 minutes,
mounted between cellulose tape and 4 cm wide Whatman
No. 1 chromatography paper, dried at 50°C and passed
through a radio-chromatogram scanner (Nuclear Chicago,
Actigraph III). Previous experiments had established
that radioactivity along the roots represents the sites
of uptake as less than 10% of the phosphate, sulfate or
chloride was translocated during the 15 minute uptake
period, the 5 minute wash and the drying time (Bowen
and Rovira, 1967).

This technique has shown that for the seminal root
system of wheat seedlings grown in soil for 14 days,
lateral roots accounted for 81% of the phosphate uptake,
79% of the sulfate uptake and 27% of the uptake of
chloride. Figure 1 shows scans of the apical 25 cm of
representative roots (40-45 cm long) which had taken up
radioactive anions. Removal of lateral roots after the
uptake and washing but before mounting on the paper,
removed all of the uptake represented by the shaded
parts of the charts.

By making use of the $\beta$ emissions from $^{65}Zn$ we have
used this technique to study the sites of uptake of

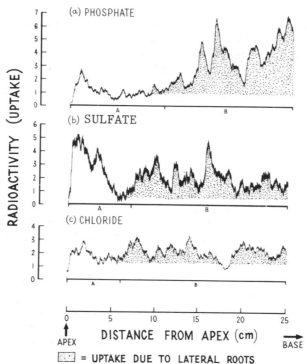

FIG. 1 Uptake of phosphate, sulfate and chloride by seminal roots of 14 day wheat
grown in soil and labelled for 15 minutes
in solution.

(a)   Phosphate from solution of $5 \times 10^{-4}$ M calcium sulfate + $5 \times 10^{-6}$ M potassium dihydrogen phosphate
with $^{32}$P-phosphate at 400 μc/l.

(b)   Sulfate from solution of $5 \times 10^{-4}$ M calcium chloride + $5 \times 10^{-6}$ M potassium sulfate with $^{35}$S-sul-
fate at 800 μc/l.

(c)   Chloride from solution of $5 \times 10^{-4}$ M calcium sulfate + $2 \times 10^{-3}$ M potassium chloride with $^{36}$Cl-
chloride at 200 μc/l.

A = Apical and prelateral zones
B = Root zone bearing lateral roots

zinc also into wheat, clover and pine roots; the uptake
of zinc is greatly increased by the presence of lateral
roots and the sites of zinc uptake in general corresponded with the sites of phosphate uptake (Rovira,
Tiller and Bowen, unpubl.).


## TRANSLOCATION OF PHOSPHATE FROM ROOTS


The results obtained by short term labelling revealed the tremendous increase in phosphate uptake following lateral root production.  In order to assess the
value of this uptake to the nutrition of the above
ground parts of the plant it is necessary to measure
the translocation of this phosphate from the different
zones of the root.  This was measured by labelling a
large number of plants with $^{32}$P-phosphate in $5 \times 10^{-6}$ M
potassium dihydrogen phosphate and $5 \times 10^{-4}$ m calcium
sulfate for 15 minutes, harvesting some immediately,
transferring the remainder to non-radioactive potassium
dihydrogen phosphate-calcium sulfate solution and after
one hour, scanning roots and tops of these plants and
calculating the decline in $^{32}$P in various parts of the
roots during this interval.  We have shown that proportionally more phosphate was translocated from the lateral root zone than from the apical region (Fig. 2).

Russell and Sanderson (1967) in a study with older
non-sterile barley roots using potometers, through
which $^{32}$P-phosphate was applied to small zones of roots,
also found that the apex of the seminal primary root
retained more phosphate than the zone behind the apex;
however, they found proportionally less phosphate
translocated from individual lateral roots than from
the main axis and from nodal roots.  These results with
lateral roots are at a variance with our results –
probably due in part to differences in experimental
techniques, species and age of plant, and nutrition
before and during the uptake experiments.

BOWEN AND ROVIRA

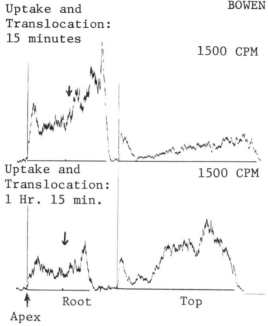

Uptake and
Translocation:
15 minutes

1500 CPM

Uptake and
Translocation:
1 Hr. 15 min.

1500 CPM

Root          Top

Apex

FIG.2  Translocation of phosphate from roots of
wheat grown in nutrient solution for 5
days, pulse labelled with $^{32}$P-phosphate for 15
minutes in $5 \times 10^{-6}$ M potassium dihydrogen phosphate and $5 \times 10^{-4}$ M calcium sulfate and then
transferred to $^{31}$P-phosphate for 1 hour.  Only
one of the three long roots of each wheat plant
is shown for each scan.  Roots and tops were
scanned at 1500 cpm full scale deflection with
1.5 mm slit, ↓ Beginning of lateral root zone.

This technique was also used with wheat
plants grown in soil at 20°C for 14 days; the
seminal roots averaged 32 cm with 20 to 30 laterals per root and showed a high translocation
of phosphate from the lateràl root zone after
1 hour 15 minutes.

EFFECT OF SOIL TEMPERATURE AND WATER ON ROOT GROWTH
AND MORPHOLOGY

During some of the studies on the uptake of phosphate described above we observed that the patterns of
uptake were affected by glasshouse conditions under
which the plants were grown.  This observation led to
experiments in which plants were grown in a Red Brown
Earth at root temperatures of 10°C and 20°C and uptake
conducted at these temperatures.

Soil temperature exerted a marked effect upon the
growth and morphology of wheat roots and Table 1 shows
that the growth of the primary seminal roots and primary
lateral roots of wheat is greatly reduced at 10°C.  Thus,
at low soil temperatures the growth of wheat may be limited by its inability to explore soil for less mobile
nutrients such as phosphate.

TABLE I

Effect of Temperature and Water Content of Soil on
(a)   the Length (cm) of the Main Axis of Seminal Roots
and (b)   the Number of Lateral Roots/Axis of Wheat at
14 Days

| Water Content | Temperature of Soil (°C) | | | |
| of Soil (%) | 10 | | 20 | |
| | a | b | a | b |
|---|---|---|---|---|
| 10 | 16 | 5 | 27 | 82 |
| 16 | 16 | 3 | 38 | 57 |

Soil water content affected root growth and morphology and hence the utilization of nutrients from
both soil reserves and applied fertilizer by roots will
be influenced by soil water content.  Although the main
axis of the primary seminal roots was reduced at the
lower water content, the number and total length of
primary lateral roots were increased; this resulted in
a greater total length of root per plant.  Another observation we made which has implications in the utilization of nutrients with low diffusivity through soil,
e.g. phosphate, was that at 10 per cent water, root
hairs were more abundant, longer and were not decom-
posed as rapidly as root hairs at 16 per cent water.

UPTAKE AND TRANSLOCATION OF PHOSPHATE AT 10°C and 20°C

Roots grown at 10°C showed a low uptake in the
apical regions with an almost uniform uptake for the
remainder of the root.  By contrast, roots grown at
20°C with uptake at 20°C showed a high peak of uptake
in the apical 3 cm, a low uptake behind the apex, and
a very high uptake in the lateral root zone (Fig. 3).
Bowen (1970) has found that roots of *Pinus radiata*
showed patterns of uptake similar to wheat roots at
low and high temperatures.  These striking differences
in patterns of uptake due to soil temperature indicate
that our concepts of uptake along roots in models of
soil-plant-nutrient systems must be adjusted to accommodate the influence of temperature.

Temperature affected translocation markedly, on
raising the temperature of the uptake solution from
10°C to 20°C, with all tops at 20°C, translocation was
trebled with roots grown at 10°C, and more than quad-
rupled from roots grown at 20°C.

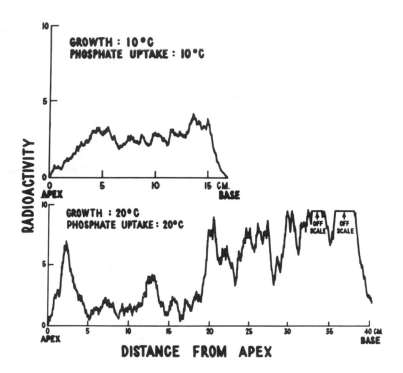

FIG. 3   Sites of phosphate uptake along wheat
         roots grown in soil at 10°C and 20°C
         with uptake from solution for 15 min-
         utes at 10°C and 20°C respectively.


CONCLUSIONS


Integration of our results with the concepts of
ion uptake from soil (Nye, 1966; Nye and Tinker, 1969)
highlight the significance of lateral root production
in (i) greatly increasing the total length of actively
absorbing root which thus increases the volume of soil

being exploited for nutrients, (ii) maintaining a high
uptake rate and thus maintaining a high concentration
gradient for diffusion of ions from soil to root over
relatively large parts of the root and (iii) transporting a high proportion of absorbed phosphate to the
tops. Unfortunately, little is known about the genetic
and environmental factors affecting the initiation and
growth of lateral roots. If we wish to improve the
utilization of both fertilizer and soil phosphate by
our domestic plants in southern Australia we should
select and/or breed varieties which will grow faster
and produce more lateral roots at soil temperatures of
10°C or less than do our present varieties.

REFERENCES

1.  BOWEN, G.D. (1968)  Chloride efflux along *Pinus
    radiata* roots.  *Nature 218*: 686-687.

2.  _____ (1969)  The uptake of orthophosphate
    and its incorporation into organic phosphates
    along roots of *Pinus radiata*.  *Aust. J. Biol.
    Sci. 22*: 1125-36.

3.  _____ (1970)  Effects of soil temperature on
    root growth and on phosphate uptake along *Pinus
    radiata* roots.  *Aust. J. Soil Res. 8*: 31-42.

4.  BOWEN, G.D. and ROVIRA, A.D. (1966)  Microbial fac-
    tor in short-term phosphate uptake studies with
    plant roots.  *Nature 211*: 665-666.

5.  _____ (1967)  Phosphate up-
    take along attached and excised wheat roots
    measured by an automatic scanning method.  *Aust.
    J. Biol. Sci. 20*: 369-378.

6.  BOWEN, G.D. and ROVIRA, A.D. (1969) New techniques to study nutrient relations in plants. *Atomic Energy in Australia 12*: 2-7..

7.  NYE, P.N. (1966) The effect of the nutrient intensity and buffering power of a soil, and the absorbing power, size and root hairs of a root, on nutrient absorption by diffusion. *Plant Soil 25*: 81-105.

8.  NYE, P.H. and TINKER, P.B. (1969) The concept of a root demand coefficient. *J. Appl. Ecol. 6*: 293-300.

9.  ROVIRA, A.D. (1969) Diffusion of carbon compounds away from wheat roots. *Aust. J. Biol. Sci. 22*: 1287-90.

10. ROVIRA, A.D. and BOWEN, G.D. (1968 a) Anion uptake by plant roots: Distribution of anions and effects of micro-organisms. *9th Inter. Congr. Soil Sci. Trans.* 2: 209-217.

11. _____ (1968 b) Anion uptake by the apical region of seminal wheat roots. *Nature 218*: 685-686.

12. RUSSELL, R.S. and SANDERSON, J. (1967) Nutrient uptake by different parts of the intact roots of plants. *J. Exp. Bot. 18*: 491-508.

*Questions to Mr. Bowen*

CLEMENTS:  Did you restrict absorption to the very oldest
part of some roots to determine whether they had adsorption there too in the roots, somewhat paralleling the
work done at Duke with water?

BOWEN:  The oldest roots we have used in studies reported here have been three week old pine roots and 14
days for wheat roots.

    Previously I have taken mycorrhizal roots from
established pine trees and found that in the older parts
uptake was slow.  Much of this appeared to be an adsorption phenomenon, which rather surprised me.

STEWARD:  I have two questions, both quite brief.  I
gather that what you were plotting was absolute amounts
of radioactivity along root.  Have you put these on a
concentration basis in the cells of the root?

    The second point is: have you ever worked with a
root that never branches, like narcissus?

BOWEN:  Actual uptake is readily obtained from the accumulated radioactivity.  We have not examined uptake
on a per cell basis yet nor worked with plants whose
roots do not branch.  At present we are examining mechanisms of uptake along the roots, and some phosphate
analyses of different parts of the root have been made.

POULSEN:  I would like to ask how you would test your
uptake.  Is it on the basic dry weight of your root, or
the fresh root weight?

BOWEN:  With this technique it is difficult to obtain a
fresh weight or a dry weight on parts of the root after
scanning because you have mounted it.

    We examine uptake per centimeter of root and, by

measuring the radius of the root, we calculate this uptake per surface area; and it seems to me that this is
the kind of thing that the plant is interested in.

POSSINGHAM:  Have you ever done these experiments and
then transferred the plants to water?  How much of what
you record is uptake in free space, and how much of it
is incorporation?

BOWEN:  We have examined the loss of phosphate from the
root to calcium sulphate solution, and we find that in a
period of three hours our loss of radioactive phosphate
to the surrounding solution is between 3% and 5%, and
from all portions of the above.  We think, therefore,
that any such loss does not distort our translocation
data, which is obtained from "pulse chase" experiments.

    What we measure is uptake and not exchange.  Also
uptake at 2°, or in the presence of potassium cyanide
is eliminated.  We examine biochemical incorporation by
doing our initial scan, then extracting readily acid
soluble inorganic phosphates by freezing and thawing in
0.1N perchloric acid which extracts all of these phosphates, but leaves the skeleton of the root.

    We then scan the root, then remove nucleic acid
fractions, rescan the same root and by differences, we
can calculate incorporation into the readily acid-soluble
phosphate, and the nucleic acid fraction and the residual
phospholipid-phosphoprotein at each portion of the root.

# MICROBIAL EFFECTS ON NUTRIENT UPTAKE BY PLANTS

Albert D. Rovira and Glynn D. Bowen

*CSIRO, Division of Soils, Glen Osmond, South Australia, Australia*

ABSTRACT

Non-infective rhizosphere micro-organisms, ecto-trophic mycorrhizas of forest trees and endotrophic my-corrhizas of horticultural, agricultural, pastoral and forestry crops are examined for their effects on plant characteristics important in ion uptake from soil. Models of ion uptake will have to accommodate large effects caused by micro-organisms, viz. microbial absorption of phosphate, modification of root hair length, radiation of fungal hyphae into soil, changes in root length and morphology, and microbial effects on plant metabolism.

Plant growth responses to ectotrophic and endo-trophic mycorrhizas are examined in relation to the present concepts of ion uptake from soil. Hyphal growth into soil and longevity in uptake are seen as major factors in plant response and in differences between fungi in their stimulation of nutrient uptake. These factors increase effective volumes of soil for phosphate with ectotrophic mycorrhizas of pine, where large inter-root distances occur. Field responses to inoculation with selected strains of mycorrhizal fungi are related to occurrence and distribution of naturally occurring strains and their relative efficiency in plant stimulation.

307

## INTRODUCTION

Plant roots support large numbers of microorganisms and often the roots are infected by symbiotic fungi ('mycorrhizas'). In this paper we examine the effects of three quite different associations on ion uptake by plants, their effects on plant parameters important in ion uptake from soils (Nye, 1966; Olsen and Kemper, 1968) and comment on field application of these studies. A more detailed treatment of some aspects of these topics has been given by Bowen and Rovira (1969).

## NON-INFECTING RHIZOSPHERE MICRO-ORGANISMS

The large diverse microbial population in the rhizosphere usually develops to within 1 cm of the apex with even rapidly growing roots and forms a well-marked but discontinuous compartment between roots and soil; in most situations 75 to 90% of the root surface is covered. Often the micro-organisms are embedded in 'mucigel' of plant origin (Jenny and Grossenbacher, 1963) but they also produce mucilage which may enhance soil-plant contact.

Soil micro-organisms affect availability of in-organic nutrients by diminishing or increasing solubility or by chelate production. Reported increases in plant growth following inoculation with phosphate dissolving bacteria ('phosphobacterin') have been equivocal (Mishustin and Naumova, 1962) and, when they do occur, as with *Azotobacter* inoculation (Rovira, 1965), they are probably due to growth factors rather than phosphate solution or nitrogen fixation. We envisage non-infecting micro-organisms of the rhizosphere may

also affect at least four root parameters important in
ion uptake:

## (i)  *Concentration of ions at the root surface*

        Ions pass through the external microbial compart-
ment before reaching the root and the extent of depletion of ions in solution will depend on the concentration in soil solution, rate of flow, and microbial
absorption rates. We have found (Bowen and Rovira 1966,
Rovira and Bowen 1966, Bowen 1969) that uptake of phosphate from 5µM phosphate over 15 minutes by non-sterile
pine, tomato, wheat and clover plants was 50 to 324
percent higher than uptake by sterile plants. Barber
(1966) and Barber and Loughman (1967) using barley over
24 hours, also found higher phosphate uptake by non-
sterile roots. Although no quantitative separation of
phosphate absorbed into roots from that into micro-
organisms on the root has yet been achieved, the increase in phosphate incorporation into the nucleic
acid fraction with non-sterile plants (Rovira and Bowen
1966, Barber and Loughman 1967, Bowen 1969) indicates
that much of the phosphate is absorbed by the rhizosphere
microflora. This microbial capture of phosphate could be
important in decreasing phosphate supply to roots in soil
low in available phosphate as well as modifying concentration of phosphate (and possibly other nutrients) at
the root surface. Thus, conclusions on phosphate uptake
and incorporation by plants may be invalid unless roots
are completely free of micro-organisms - a precaution
not often observed. The situation with ions other than
phosphate awaits investigation although Epstein (1968)
has found some effects of micro-organisms on uptake of
potassium by excised roots.

(ii)  *Root hair growth*


Studies on subterranean clover in sand culture
(Bowen and Rovira 1961) have shown that a number of soil
micro-organisms can drastically reduce root hair growth.
In soil where phosphate uptake is largely confined to
the root "cylinder" defined by root hairs (Lewis and
Quirk 1965) the reduction in root hair length from 200
to 100µm on a root radius of 250µm found by Bowen and
Rovira (1961) would reduce effective soil volumes for
phosphate by approximately 40 percent.


(iii)  *Root length and morphology*


Root growth is important in the uptake of nutrients (e.g: Olsen and Kemper 1968) and production of
lateral roots increases the uptake ability of roots
(Bowen and Rovira 1970).  Bowen and Rovira (1961)
showed that in sand a general soil microflora reduced
root growth of subterranean clover, tomato, phalaris,
and pine seedlings by 25–52 per cent and lateral root
production by 20–53 per cent.  The reduction of root
growth and root hair length could lead to 54–64 per
cent reduction in phosphate uptake.  Fortunately, not
all micro-organisms have these effects and we have isolated organisms which stimulate root growth, as have
Domsch (1963) and Otto (1965).  Rovira (unpublished)
has found that *Azotobacter chroococcum* increases the
growth of roots and root hairs of wheat in nutrient
solution, further study of such stimulation may result
in techniques by which the nutrient uptake by plants
from soil is increased.

(iv)  *Uptake ability and metabolism of plants*


     Rhizosphere micro-organisms change the metabolism
of plants, e.g. the transport of phosphate to tops of
non-sterile tomato plants may be four times that of
sterile (Bowen and Rovira 1966) and occasionally a
greater incorporation of phosphate into nucleic acid
fractions of plant tops occurs with non-sterile plants
(Rovira and Bowen 1966).  Inoculation of plants with
*Azotobacter* and a number of other micro-organisms often
induces early flowering (Rovira 1963, Brown *et al.*, 1964).
The reasons for such effects (and others) need more study
but they are consistent with effects of plant growth
regulators known to be produced by soil micro-organisms.


## ECTOTROPHIC MYCORRHIZAS


     The stimulation of uptake from soils of low nutri-
ent status by ectotrophic mycorrhizas on a number of
tree species is well known (see Harley 1969).  We will
deal only with nutrient uptake in *soil* and the possibility of selection of highly efficient organisms in
this regard.  Although we refer particularly to phosphate, mycorrhizas increase uptake of many nutrients
and probably also act as water uptake organs.  Taking
the parameters used above, integration of the ectotrophic mycorrhizal association into existing models
for ion uptake must recognize:


(i)  *Concentrations of ions at root surface*


     The external microbial compartment around the in-
fected short lateral roots is usually well-developed,
compact, and can be several hyphae wide.  As this

compartment has a high affinity for phosphate (see Harley 1969), under normal soil conditions and with a well-developed fungus sheath, the surface of the root contacts very little phosphate, the fungal component absorbing it all and transporting it between the cortical cells of the root up to the endodermis. The fungi absorb a wide range of organic compounds (thus short-circuiting classical mineralization cycles) and the fungal sheath can act as a storage organ for phosphate. Bowen and Theodorou (1967) reported three-fold differences in phosphate uptake from solution culture between mycorrhiza types with up to 4½ times the uptake of uninfected roots. The large differences in uptake by different mycorrhiza types from solution may be less significant in soil where phosphate uptake is largely diffusion controlled.

(ii)  *Root hair formation*


Root hairs are suppressed with ectotrophic mycorrhizas but are often replaced by fungal growth into soil (Hatch 1937). By analogy with the importance of root hair length in phosphate uptake, Bowen (1968) considered fungal growth into soil to be a major factor in increasing effective soil volume for phosphate uptake - inter-root distances for pine being large compared with radii of diffusion of phosphate to roots. The extent of fungal growth into soil varies greatly between fungi (0-3.5 cm) and different plant responses to different fungi may be related to the extent of fungal growth from the root.

(iii)  *Root morphology*


Root radius is increased by production of an enveloping fungal sheath, by radial elongation of cortical cells and by branching.


(iv)  *Root metabolism*


Translocation of ions and plant metabolism is probably affected by production of auxins (Slankis 1958) and cytokinins (Miller 1967) by the fungi. These may contribute also to the pronounced *longevity* of ectotrophic mycorrhizas (often of several months to a year) which increases the use of soil between the relatively widely spaced roots by increasing radii of diffusion and convection of ions and by constantly removing phosphate in the soil solution in equilibrium with poorly soluble organic and inorganic phosphates (Bowen and Theodorou 1967).

Although responses to inoculation should be maximal in the absence of naturally occurring mycorrhizal fungi, plant growth stimulation from inoculation can occur where mycorrhizal fungi are low in number, unevenly distributed, or poor in efficiency. In field soils containing ectotrophic mycorrhizal fungi Theodorou and Bowen (1970) found that at 32 months inoculated trees of *Pinus radiata* were 146 cm high and non-inoculated trees 108 cm; different strains of mycorrhizal fungi varied in their effects on growth of the host.


ENDOTROPHIC MYCORRHIZAS


Vesicular-arbuscular ("v.a.") mycorrhizas, caused by species of the phycomycete *Endogone* are potentially the most important non-nitrogen fixing symbioses with plants. This association occurs on a vast range of horticultural, agricultural, pastoral and forestry plants, and under conditions of low nutrient availabilit

can increase nutrient uptake and plant growth (see Bowen
and Rovira 1969).   Large increases in plant growth and
uptake of P, K, Ca, Mg and trace elements have been obtained on plants such as apple, maize, tobacco, tomato,
vines, onion, clover and Sudan grass (Gray and Gerdemann
1967, Daft and Nicolson 1966).

In models of ion uptake from soil, endotrophic mycorrhizas may affect:

(i)   *Concentrations of ions at the root surface*

The relative sparseness of external hyphae of v.a.
mycorrhizas suggests they do not form a physical barrier
between ions and roots, although of course the non-infective rhizosphere compartment still exists.

(ii)   *Growth of fungi into soil*

Bowen and Mosse (unpubl.) found that the mycelium
absorbed phosphate readily and translocated it to extensive growth inside the root.   Mosse (1963) observed
mycelial growth up to 1 cm from the root, this mycelial
growth into soil extends the effective soil volume for
ions such as phosphate.

(iii)   *Root morphology*

Mosse (1963) found that infected roots of clover
were longer than uninfected roots under laboratory conditions (21.6 mm vs. 13.1 mm) and produced lateral roots
more frequently (16.3 per cent of rootlets with laterals vs. 1.3 per cent).

(iv) *Effects on host metabolism*

Changes in plant metabolism probably occur and one indication of this is an increase in host cell nucleus diameter before and during arbuscule formation. Bowen and Mosse (see Bowen and Rovira 1969) found by micro-autoradiography of root sections, that a 146 per cent increase in phosphate uptake over 30 minutes by mycorrhizal clover roots was primarily due to uptake by the fungus and translocation to the vesicle and arbuscule (from whence it would move into the host via the greatly increased surface area for exchange of solutes and metabolites afforded by arbuscules). There appeared to be no major effect of the fungus in stimulating uptake by the plant cell; however as some obligate parasites increase auxins and cytokinins in plants (Linskins 1968) similar phenomena may occur with v.a. mycorrhizas.

Considering that large plant responses to inoculation have been obtained in the absence of naturally occurring *Endogone* spores, what are the prospects of stimulating plant growth by inoculation with *Endogone* under commercial field conditions? The prospects are good as shown by Mosse, Hayman and Ide (1969) who increased the growth of *Liquidambar styraciflua* by 65 per cent and of onion by 140 per cent with inoculations into the soil containing a natural population of *Endogone*.

CONCLUSIONS

Micro-organisms significantly modify ion uptake parameters from soil and as such have to be superimposed on existing models, not as special cases but as the normal case. Effects of micro-organisms are not always in the same direction and raise the interesting question of the definition of a pathogen when ion uptake and plant growth are decreased by non-infecting soil organisms.

Soil microbiologists have a real role to play in controlling the activities of micro-organisms in ion uptake but much more must be learnt about population ecology of the rhizosphere before we can fully control the non-infecting rhizosphere organisms. By their infective nature, the mycorrhizal fungi offer more immediate prospects of manipulation, and indeed such study is well advanced in ectotrophic mycorrhizas of trees and the endotrophic mycorrhizas of other plants of economic importance.

REFERENCES

1.   BARBER, D.A. (1966)  Effect of microorganisms on nutrient absorption by plants. *Nature 212*: 638-40.

2.   BARBER, D.A. and LOUGHMAN, B.C. (1967)  Effect of microorganisms on absorption of inorganic nutrients by intact plants.  II: Uptake and utilization of phosphate by barley plants grown under sterile and non sterile conditions. *J. Exp. Botany 18*: 170-76.

3.   BOWEN, G.D. (1968)  Phosphate uptake by mycorrhizas and uninfected roots of *Pinus radiata* in relation to root distribution. *Proc. 9th Int. Cong. Soil Sci. Adelaide 2*: 219-28.

4.   ——————— (1969)  The uptake of orthophosphate and its incorporation into organic phosphates along roots of *Pinus radiata. Aust. J. Biol. Sci. 22*: 1125-35.

5.   BOWEN, G.D. and ROVIRA, A.D. (1961)  The effects
     of microorganisms on plant growth.  1: Develop-
     ment of root hairs in sand and agar.  *Plant
     Soil 25*: 166-88.

6.   ———————————————————— (1966)  Microbial
     factor in short term phosphate uptake studies
     with plant roots.  *Nature 211*: 665-66.

7.   ———————————————————— (1969)  The influence
     of microorganisms on growth and metabolism of
     plant roots.  In: *Root Growth,* Proc. 15th
     Easter School, Nottingham, 1968.  Whittington,
     W.J. (ed.)  Publ. Butterworths, London, pp.
     170-201.

8.   ———————————————————— (1970)  Relationship
     between root morphology and nutrient uptake.
     *Proc. 6th Int. Coll. Plant Anal. Fertilizer
     Probl.*, Tel Aviv, Israel, 1970.

9.   BOWEN, G.D. and THEODOROU, C. (1967)  Studies on
     phosphate uptake by mycorrhizas.  *Proc. 14th
     I.U.F.R.O. Cong. Munich 5*: 116-138.

10.  BROWN, M.E., BURLINGHAM, S.K. and JACKSON, R.M.
     (1964)  Studies on Azotobacter species in soil.
     III: Effects of artificial inoculation on crop
     yield.  *Plant.Soil 20*: 194-214.

11.  DAFT, M.S. and NICHOLSON, T.H. (1966)  Effect of
     Endogone mycorrhiza on plant growth.  *New Phytol.
     65*: 343-50.

12.  DOMSCH, K.H. (1963)  Der einfluss saprophytischer
     bodenpilze auf die jugendentwicklung höherer
     pflanzen.  *Pflanzenkrankh. Pflanzenschutz 70*:
     470-75.

13.  EPSTEIN, E. (1968)  Microorganisms and ion absorp-
     tion by roots.  *Experimentia 24*: 616-17.

14.  GRAY, L.E. and GERDEMANN, J.W. (1967)  Influence
     of vesicular-arbuscular mycorrhizas on the up-
     take of $P^{32}$ by *Liriodendron tulipifera* and
     *Liquidambar styraciflua*. *Nature 213*: 106-7.

15.  HARLEY, J.L. (1969)  *Biology of Mycorrhiza*, 2nd
     Ed., Leonard Hill, London, p. 334.

16.  HATCH, A.B. (1937)  The physical basis of myco-
     trophy in the genus *Pinus*. *Black Rock For.
     Bull. 6*: p. 168.

17.  JENNY, H. and GROSSENBACHER, K. (1963)  Root and
     soil boundary zones as seen by the electron
     microscope. *Soil Sci. Soc. Amer. Proc. 27*:
     273-7.

18.  LEWIS, D.G. and QUIRK, J.P. (1965)  Phosphate dif-
     fusion in soil and uptake by plants.  III: $P^{31}$
     movement and uptake by plants as indicated by
     $P^{32}$ radiography. *Plant Soil 26*: 445-55.

19.  LINSKINS, H.S. (1968)  Host-pathogen interaction
     as a special case of interrelations between or-
     ganisms. *Neth. N.J. Pl. Path. 74*: suppl. 1-8.

20.  MILLER, C.O. (1967)  Zeatin and zeatin riboside
     from mycorrhizal fungus. *Science 157*: 1055-7.

21.  MISHUSTIN, E.M. and NAUMOVA, A.M. (1962)  Bacterial
     fertilizers, their effectiveness and mode of
     action. *Mikrobiologia 31*: 442-52 (Engl. Transl.).

22.  MOSSE, B. (1963)  Vesicular-arbuscular mycorrhiza:
     an extreme form of fungal adaptation. *Symbiotic
     Associations*, Nutman, P.S. and Mosse, B. (eds.).
     Camb. Univ. Press, pp.147-70.

23.  MOSSE, B., HAYMAN, D.S. and IDE, G.J. (1969)  Growth
     responses of plants in unsterilized soil to in-
     oculation with vesicular-arbuscular mycorrhiza.
     *Nature 224*: 1031-2.

24.  NYE, P.H. (1966)  The effect of the nutrient in-
     tensity and buffering power of soil and the ab-
     sorbing power, size, and root hairs of a root on
     nutrient absorption by diffusion.  *Plant Soil
     25*: 81-105.

25.  OLSEN, S.R. and KEMPER, W.D. (1969)  Movement of
     nutrients to plant roots.  *Adv. Agron. 20*:
     91-151.

26.  OTTO, G. (1965)  The influence exerted by fungi in
     the rhizosphere on the density and ramification
     of cucumber roots.  *Plant Microbes Relation-
     ships*, Macura, J. and Vancura, V. (eds.).  Publ.
     Czech. Acad. Sci., Prague, pp. 209-19.

27.  ROVIRA, A.D. (1963)  Microbial inoculation of crop
     plants.  1:  Establishment of free living nitro-
     gen fixing bacteria in the rhizosphere and their
     effects on maize, tomato and wheat.  *Plant Soil
     19*: 304-14.

28.  ─────────── (1965)  Effects of *Azotobacter, Bacil-
     lus* and *Clostridium* on the growth of wheat.
     *Plant Microbes Relationships*, Macura, J. and
     Vancura, V. (eds.).  Publ. Czech. Acad. Sci.,
     Prague, pp. 195-200.

29.  ROVIRA, A.D. and BOWEN, G.D. (1966)  Phosphate in-
     corporation by sterile and non-sterile plant
     roots.  *Aust. J. Biol. Sci. 19*: 1167-9.

30.   SLANKIS, V. (1958)   The role of auxin and other
      exudates in mycorrhizal symbiosis of forest
      trees. *Physiology of Forest Trees.*  Ronald
      Press, N.Y., pp. 427-43.

31.   THEODOROU, C. and BOWEN, G.D. (1970)  Mycorrhizal
      responses of radiata pines in glasshouse and
      field experiments with different fungi.  *Aust.
      Forestry 34* In Press.

*Questions to Mr. Bowen*

REDLICH:  Do you find mycorrhizas increase uptake of
only phosphorus?

BOWEN:  Many studies have been made with phosphate.
There is evidence of increased uptake of many other nutrients as well; e.g. with apple cuttings, 9 months old
showed increases in potassium, calcium and magnesium and
iron uptake with endotrophic mycorrhizas.

JACKMAN:  Do endotrophic fungi also increase the volume
of soil from which the plant can effectively feed, as
do ectotrophic mycorrhizas?

BOWEN:  This hasn't been studied very well yet, but it
seems logical to me that they do.  They can probably
use forms of phosphate other than inorganic.  We have
measured their growth out into the soil to a distance
up to a centimeter or so.

# ISOLATION AND CONCENTRATION OF TRACE ELEMENTS IN PLANT TISSUES BY ION-EXCHANGE

Octavio Carpena, Antonio Leon,* Santiago Llorente and Carlos Alcaraz

*Department of Agricultural Chemistry, University of Murcia, Murcia, Spain*

ABSTRACT

Nutrition of plant tissues was studied by ion-exchange determination of trace elements.

We investigated several causes of error, including working conditions, in determining zinc by atomic absorption spectrophotometry.

Columns of Dowex 1 × 8 (100-200 mesh) anion exchange resins were used and favorable conditions for zinc elution investigated.

An ion-exchange technique for isolating Fe, Mn, Zn and Cu is presented and compared with determinations by atomic absorption spectrophotometry. This technique is used on acid plant extracts but may also be used on soil extracts. Results compare favorably with those of other techniques.

---

*Presented this paper at the Colloquium.

INTRODUCTION


Since Lure and Filippova (1947) carried out their
experiments on the concentration of trace elements by
synthetic exchange resins, the methods of isolation and
concentration of inorganic ions have changed greatly.

At present, the anion exchange resins provide additional possibilities in plant analysis.

Formation of certain complex compounds with metallic ions in HCl solution is an important factor in
analytical separation, since the chloride complexes are
negatively charged, and are therefore fixed by the anionic
exchange resins.

Adsorption of the complexes depends on the metallic
complex itself as well as on the HCl molarity. The important studies of Kraus and co-workers (1950, 1953,
1954 and 1956), provided the basis for application of
anionic exchange resins to isolation of metallic ions.
They indicate the need to increase the HCl molarity in
order to obtain a suitable adsorption of divalent ion
complexes. The decreasing stabilities of the complexes
in the order of Zn > Fe > Cu > Mn enables a selective
elution. Kraus and Moore (1953) studied adsorption-
isolation, and developed a method for the Fe, Mn, Co,
Ni, Cu, and Zn separation, eluting with different HCl
concentrations.

Several systematic techniques for zinc separation
have been presented by a number of workers. Miller and
Hunter (1954) isolated Zn from other elements by means
of Amberlite I.R.A.-400 resins. Rush and Yoe (1954)
isolated Zn from several interfering elements with
Dowex-1 resins, eluating with 0.01N HCl. Kallman *et al.*
(1956) utilized 2N NaOH for the Zn elution. Pratt and
Bradford (1958) adapted the method of Kraus to the Zn
separation in soils. Yuang and Fiskell (1958) presented

a method for plant tissues with Dowex-1×4 resins. Ulrich
*et al.* (1959) proposed the utilization of the N/10
NaNO₃ for Zn elution followed by spectrophotometric determination. Hunter and Coleman (1960) adapted the Kraus
technique to cation spectrophotometry determinations in
plants. Recently, Carlson *et al.* (1968) applied cation
separations to plant tissues, and determined their concentrations by means of polarographic techniques.

However, these techniques are not generally accepted
by agricultural chemists, especially for Zn determination,
due to the inaccurate separations obtained using dilute
HCl. For this reason Margerum and Santacana (1960) pre-
ferred extraction with ditizone in CCl₄.

Certainly, the Zn separation utilizing .01M or .005M
HCl as eluating agents is not satisfactory. In this paper
we propose on the basis of our results different concentration ranges.


MATERIAL AND METHODS


*Columns.* The columns are of Pyrex glass, 1 cm diameter
and 25 cm high.

*Resins.* Dowex-1×8, 100-200 mesh. 6 g of dry resins
are soaked in demineralized water during 24-
36 hours.

*Analytical.* HCl analytical reagent (A.R.) KCl (A.R.)
NaNO₃ (A.R.).

*Preparation of the columns.*
1. A bed of glass wool, 1 cm thick is prepared;
2. a resin suspension is added slowly to a
height of 15 cm;
3. a stopper of glass wool is used to regulate the rate of the flow to ± 2 ml/min;

4. the columns are equilibrated with 20 ml
   9M HCl.

*Mineralization of plant tissues*

The mineralization of plant tissues is carried
out according to the method proposed by Carpena, Leon,
and Alcaraz (1968), and the mineral residues diluted
with 9M HCl.

*Instruments*

For the analytical determinations we employed the
following instruments: Beckman DB spectrophotometer -
atomic absorption accessory Beckman no. 130100, with
laminar flux burner and hollow cathode lamps for Mn,
Cu, Fe, and Zn determinations. Alternating tension
stabilizer Philips 7776/06.

*Analytical techniques*

The four trace elements studied here were determined by atomic absorption spectrophotometry. The
"Flame Notes" by Beckman Instrument Inc. were followed
except for the Zn determination, since difficulties
appeared with the recommended method. We therefore
studied the influence of lamp current variations, as
well as the air and acetylene fluxes, that are not
mentioned in the publications of David (1958) and
Ramirez-Muñoz (1967).

RESULTS

a) *Systematic technique for Fe, Mn, Cu, and Zn separation*

Ulrich *et al.* (1959), Mazoyer (1961) and Dartigues

(1966), utilized 1N NaNO$_3$ only for Zn separation. This suggested to us that we try the use of this reagent for the Mn-Cu-Fe-Zn separation.  Iron was separated with 0.5N HCl adding 1N KCl.

    Procedure:

1. After equilibrating the columns, pass through 50 ml of demineralized water and 10 ml of 9M HCl.
2. Add the sample in the 9M chloride extract and then 5 ml of 9M HCl.
3. Mn elution is achieved by 15 ml 6M HCl. The percolate is evaporated, and the residue is dissolved with 25 ml demineralized water.
4. Cu elution is carried out with 15 ml 2.5M HCl and the eluate treated as above.
5. Fe is eluted by 20 ml 1M KCl. The percolate is collected in a 25 ml flask, and made up to volume with demineralized water. The determination is carried out on 3 ml of this dissolution
6. Zn is eluted with 20 ml 0.1M NaNO$_3$. The percolate is collected in a 50 ml flask, and brought up to volume with demineralized water.

    Fig. 1 shows the elution curves obtained by the systematic technique described above. It is important to observe the marked peak, in the Zn elution curve, in contrast to that obtained when using dilute HCl.


b) *Reproducibility of the results*

    In order to test the reproducibility of the results, we used Marsh grapefruit leaves. We digested 10 g of dry leaf powder. The direst residue was dissolved in 25 ml. of HCl. In Table 1 we are showing the results obtained when ten consecutive 2 ml samples of the same preparation were passed through the column.

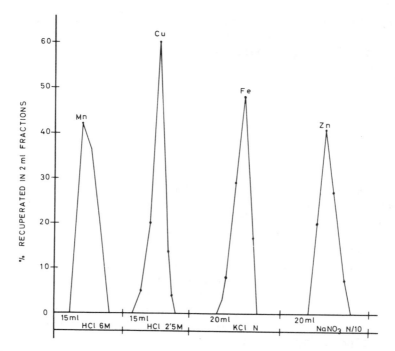

FIG. 1 Elution curves of different mineral elements
obtained after their adsorption on an amberlite
column by means of certain elements.

c) *Recuperation*

   In a similar way we determine the recuperation
(expressed as % in Table II), of different amounts of
each trace element added to the aliquots of an extract
of some vegetal tissue.

   Table II shows the results obtained with the
manganese, copper and iron determinations. In the case
of zinc, (Table III), we compared the results obtained
with our proposed method with those, when 0.005M HCl
is employed. We observed that the first method produces

TABLE I. Reproducibility of the Results

| Element | I | II | III | IV | V | VI | VII | VIII | IX | X | Mean | Mean Error | Standard Deviation | % Relative Error | % C.V. |
|---|---|---|---|---|---|---|---|---|---|---|---|---|---|---|---|
| Mn | 24 | 23 | 23 | 25 | 24 | 23 | 23 | 24 | 23 | 23 | 23.5 | 0.22 | 0.71 | 2.12 | 3.01 |
| Cu | 9 | 9 | 9 | 8 | 9 | 10 | 9 | 9 | 9 | 8 | 8.9 | 0.58 | 0.57 | 4.49 | 6.38 |
| Fe | 70 | 69 | 69 | 70 | 69 | 72 | 70 | 71 | 70 | 69 | 69.7 | 0.42 | 1.33 | 1.35 | 1.92 |
| Zn | 32 | 30 | 31 | 32 | 32 | 32 | 32 | 31 | 33 | 32 | 31.7 | 0.26 | 0.82 | 1.83 | 2.59 |

Fraction No.

TABLE II.   Recuperation of Mn, Cu and Fe Using Resin
Column  Technique Using Consecutive 2 ml
Fractions of the Extract

| | μg. of oligoelement | | | |
| | added | obtained | recuperated | % recuperation |
|---|---|---|---|---|
| Manganese | 0 | 14.4 | – | – |
| | 6 | 20.2 | 5.8 | 97 |
| | 12 | 27.0 | 12.6 | 105 |
| | 18 | 31.8 | 17.4 | 97 |
| | 24 | 38.0 | 23.6 | 98 |
| | 30 | 44.1 | 29.7 | 99 |
| Copper | 0 | 5.7 | – | – |
| | 2 | 7.6 | 1.9 | 95 |
| | 4 | 9.4 | 3.7 | 93 |
| | 5 | 10.5 | 4.8 | 96 |
| | 6 | 11.5 | 5.8 | 97 |
| | 7 | 12.7 | 7.0 | 100 |
| Iron | 0 | 57.5 | – | – |
| | 6 | 63.3 | 5.8 | 97 |
| | 12 | 68.9 | 11.4 | 95 |
| | 18 | 74.8 | 17.3 | 96 |
| | 24 | 81.3 | 23.8 | 99 |
| | 48 | 104.2 | 46.7 | 97 |

a recovery of 95%, against 80-90%, obtained when dilute
HCl is utilized. The advantage of our method is evident,
and with this procedure, the determination is faster
and results in a smaller final volume.


d) *Zinc determination by atomic absorption spectrophotometry*

We also studied the influence of the rate of air
flow, using the same pressure of acetylene, working with

TABLE III.  Recuperation of Zn Using Two Methods of
            Elution (using consecutive 2 ml fractions
            of the extract)

| | μg. oligoelement | | |
| | added | obtained | recuperated | % recuperation |
|---|---|---|---|---|
| | 0 | 19.8 | – | – |
| | 3 | 22.7 | 2.9 | 97 |
| Zinc | 6 | 25.5 | 5.7 | 95 |
| (NaNO$_3$N/10) | 9 | 28.4 | 8.6 | 96 |
| | 12 | 31.7 | 11.9 | 99 |
| | 15 | 34.2 | 14.4 | 96 |
| | 0 | 18.9 | – | – |
| | 3 | 21.3 | 2.4 | 80 |
| Zinc | 6 | 23.7 | 4.8 | 80 |
| (HCl 0.005M) | 9 | 26.8 | 7.9 | 88 |
| | 12 | 29.5 | 10.5 | 87 |
| | 15 | 32.4 | 10.5 | 90 |

five levels of lamp current and different zinc concentrations (Table III). We observed that the %A variations due to air pressure depend on the level of the lamp current, showing considerable differences in the means of %V.C. for the four concentrations studied. Those differences were smaller for lamp current intensities of 12 and 9 mA, as compared with difference at the lamp current intensities 8 and 10 mA which had more influence than variations in the gas flow.

More directly, we studied the suitable current intensity, certifying the experimental variations in absorbance/concentration relationship for seven Zn concentrations, working with the same lamp current and gas flow levels mentioned before. Minor variations were obtained with the lamp current of 12 mA (mean % V.C. 2.5), following that of 9 mA (3.2%); and corresponding

TABLE IV.  Effect of Variation in Lamp Current and Rate
of An Flow on Determination of Zn Using
Atomic Absorption Spectrophotometry

| Lamp current mA. | ppm Zn | % V.C. of the % A | | | |
|---|---|---|---|---|---|
| | | 0.05 | 0.10 | 0.15 | 0.20 |
| 8 | | 3.9 | 4.1 | 4.1 | 3.8 |
| 9 | | 3.4 | 1.4 | 1.7 | 2.0 |
| 10 | | 6.2 | 5.0 | 3.9 | 3.3 |
| 11 | | 4.6 | 4.2 | 2.3 | 3.0 |
| 12 | | 4.3 | 2.4 | 0.8 | 1.6 |

to the greater variations in the current intensity levels
of 10 mA (5.2%) and 8 mA (5.4%).

Due to these results, we selected a lamp current of
12 mA for the zinc determination on plant and soil extracts, previously separated in the columns.


DISCUSSION


The systematic separation which we utilized offers
a great advantage for the plant and soil analyses of
trace elements, and eliminates a great number of problems in their determination. The interferences due to
the relatively higher concentrations of some cations,
such as calcium, are eliminated by this method.

Moreover, the present method has the advantages
of rapidity, accuracy and reproducibility.

We reported our results in the separation and determination of zinc. The elution curve shows a clear
peak, which is an index for its satisfactory separation. Also recovery and reproducibility experiments

show a high level of accuracy, and the variability of the method using dilute HCl is eliminated.

The above technique is utilized by us generally on plant tissues and soils extract. A high level of calcium carbonate in the soil, does not interfere with the determinations, if this method is utilized.

## REFERENCES

1.  CARLSON, R.M., GRASSI, R.L., MAZZA, C.A., SANTA-MARIA, R.M. and VALLEJOS, W.E. (1968) *Agrochimica 12*: 150-156.

2.  CARPENA, O., LEON, A. and ALCARAZ, C. (1968) *Comptes-Rendus II, Col. Eur. Med. Quim. Plant. Ult.* Sec. 3.

3.  DAVID, D.J. (1958) *Analyst 83*: 665.

4.  KALLMAN, S., STEELE, C.G. and CHU, N.J. (1956) *Anal. Chem. 28*: 230.

5.  KRAUS, K.A. and MOORE, G.E. (1953) *J. Am. Chem. Soc. 75*: 1460-62.

6.  KRAUS, K.A. (1954) *J. Phys. Chim. 58*: 11-17.

7.  KRAUS, K.A. and NELSON, F. (1956) *Proc. Int. Conf. Peaceful Uses of Atomic Energy 7*: 113-131.

8.  LUR'E, YU. YU. and FILIPPOVA, N.A. (1947) *Zavodskaya Lab. 13*: 539.

9.  MARGERUM, D.W. and SANTACANA, F. (1960) *Anal. Chem. 32*: 1157-1161.

10. MAZOYER, R. (1961) *Ann. Agron. 12*: 609-617.

11.  MILLER, C. and HUNTER, J.A. (1954) *Analyst. 79*:
     483-92.

12.  MOORE, G.E. and KRAUS, K.A. (1950) *J. Amer. Chem.
     Soc. 72*: 5792-93.

13.  PRATT, P.F. and BRADFORD, G.R. (1958) *Soil Sci.
     Soc. Amer. Proc. 22*: 399.

14.  RAMIREZ-MUNOZ, J. (1967) *Anal. Edaf. Agrob.*,
     pp. 1211-1226.

15.  RUSH, R.M. and JOE, J.H. (1954) *Anal. Chem. 26*:
     1345-47.

16.  ULRICH *et al.* (1959) *Calif. Agr. Exp. St. Bull
     766*: 71-76.

17.  YUANG, T.M. and FISKELL, J.A. (1958) *J. An. Off.
     Agr. Chem. 41*: 424-28.

# GENERAL DISCUSSION

BOWEN (Session Leader): We have to see soil chemistry in terms of transport processes and what they mean for the plant. I don't think you can really separate evaluation of nutrient potential of a soil from the plant that you are studying and from the environmental conditions occurring during growth.

A second aspect is the root parameters that are involved in uptake. Can we measure the relative importance of these? Can we modify them?

## Soil-plant transport processes

I think Dr. Barber's point that we should be examining or analysing for what affects the plant, e.g. potassium concentrations rather than $K/\sqrt{Ca}$ is a good one. Wilde, Rowell and Ogunfoura recently reported a better relation between yields (ryegrass, flax) and K concentration than $K/\sqrt{Ca}$.

In presenting your paper you showed that in some cases you get a calcium accumulation at the root and with others you get a calcium depletion, and your explanation was based on the relative abilities of different plants to absorb calcium. Wilkinson, Loneragen and

333

Quirk have shown $Ca^{45}$ depletions around the root, but
suggested that water may be moving to the roots prefer-
entially through the larger pores and this may lead to
less calcium in the root than would have been expected
otherwise.

BARBER:  We have been able to show accumulation around
the root very clearly.  Incidentally, with the small
volume of the soil-box, one has to water the soil very
carefully (we use hypodermic needles in the back cor-
ners of the box).  Otherwise water will flood across
the face of the box, removing the calcium and thus producing anomalous results.

In some of our work we were actually more interested in phosphorus, but we measured calcium because we
thought that calcium accumulation would affect phosphorus uptake as well as pH.  So we were looking at some
species and evaluating them in terms of calcium accumulation about the root and the degree to which they
changed the pH, to see if this could be related to the
ability of these species to take up phosphorus.

CLEMENTS:  The root system of sugar cane often becomes
quite inactive, especially in winter; there are then no
root tips coming out anywhere.  If, during this stage,
you apply either of the common nitrogen fertilizers,
such as ammonium sulphate or urea, then, shortly, a
very great mass of secondary roots breaks through.
Therefore I would conclude that there was absorption of
the nutrient by these 'inactive' roots.

BARBER:  We haven't studied older roots.  Most of our
work has to be with younger roots as they grow in a pot.

COÏC:  Je voudrais donner une indication, qui se
rapporte peut-être à la discussion sue le calcium et
le potassium des recines.  Dans le maïs, nous avons
trouvé une différence dans le rapport calcium sur
potassium des petites et des grosses racines, allant

de 1 à 10.  Il y avait un rapport calcium sur potassium
10 fois plus grand dans les petites racines que dans
les grosses racines en concentration de moli.

BOWEN:  Would Dr. De Wit tell us what he sees as the
advantages and disadvantages of simulation approaches
to soil-root interaction.   I would also ask whether
he has, in fact, tested the accuracy of the simulation?
The model can generate ideas but these must be experimentally tested.

DE WIT:  We have to look at mathematics and whether the
simulation is correct or not.  We can test for the movement into cells, the movement of ions and the movement
of water in soil, and in this kind of thing we are
satisfied that the simulation gives correct answers
in complex systems.

When you come to precipitation questions, I think
the simulation itself will give the correct answer, but
I am much more worried about whether the chemistry
behind it is really good chemistry.

Now we can ask ourselves what can we do with the
calculations?  E.g. we may want to grow grass with 4%
nitrogen and so much potassium in it, we can then ask
if we have the grass growing at a good rate, which is
80 kilo per hectare, with 4% nitrogen in it, with so
much potassium in it.

First, how much nutrient goes to the roots with
mass flow and diffusion of different ions?  If we
decide with a fast growing plant that mass flow and
diffusion cannot account for the whole process, we have
to count on exploration by new roots.

When we wish to follow a certain upward flux in
the plant, the question arises whether the system is
suitable for this upward flux or not, and what we have
to do to make it suitable?

*Root parameters involved in uptake*

BOWEN:  It would be interesting to calculate the relative importance of various root factors and then to
select or breed for them.

KAFKAFI:  Mr. Bowen stressed the effect of laterals.
Is this not just a new root, similar to the main one
going out and exploring new soil volume.

BOWEN:  I agree with you to a very large extent, but we
are also increasing the number of uptake points.  The
behavior of apices of laterals in translocation of phosphate appears to be different from the apices of the
main root.  Lateral root apices translocate phosphate
very effectively, and are not functionally identical
with apices of main roots.

KAFKAFI:  When the tip of the lateral is emerging, then
the phosphate at this point in the soil is depleted.

BOWEN:  That is a very fair point, but the zone of depletion of phosphate around the main root is very
narrow, often less than a millimeter, and it will not
take long for the lateral to grow through this to
exploit new soil.

DE WIT:  We should distinguish between plants.  With
onion and tulip there is little branching, and the
roots are so thick, that practically all the ions have
to move to the plant by mass flow.  You have to be very
careful with the fertilization and with water supply to
control it.

But we may possibly have a field situation with
other plants, where root branching and exploration of
new soil give all the supply, where we don't have to
rely practically on mass flow.  If you calculate surface areas of root systems, as affected by branching,
as you do with corn, you find especially at different

temperatures, how much it affects leaf area and top/
root ratio.

BARBER: How much root system do we need? Maybe we
have two or three times as much root now than we could
get along with.

We are growing corn without cultivation and with
cultivation. With no tillage, you have about one half
the aeration porosity, and because of that, the root
system is different. Where there was no tillage I had
only one half the roots, but top growth and the nutrient composition in the top was the same with and
without tillage. Maybe we have far more roots now than
we need?

DE WIT: If you do a number of experiments on ploughing
with different crops, you often see a different distribution of roots of different densities, but in general,
I would say that you do not have half the number of
roots. We have observed 20% differences in crops between non-ploughing and ploughing.

On the other hand, I want to say that I think that
we can grow good crops without ploughing with most
plants.

HERSHENSON: Mr. Bowen, you told us of differences between several strains of the fungi concerning the uptake of phosphate. Would you attribute it more to the
respective surfaces and to the quantity of the special
strain, or specifically to the relative uptake of the
phosphate by different strains?

BOWEN: There is no doubt that different mycorrhizal
fungi show differences in uptake from solution culture.
What really worries me is the tendency to extrapolate
from uptake in solution culture to uptake in the soil
situation. We have to place this uptake work into
the context of root distribution and the movement of

ions to roots. From what we know of phosphate, it
would seem that you can increase phosphate uptake from
soil rather more by increasing the effective volume of
the soil via selecting fungi capable of radiating appreciable distances into the soil than by selection for
high efficiency uptake from solution culture, although
this has some importance.

COÏC: Je pense que lorsque l'on parle des racines, on
pense trop exclusivement à l'absorption de l'eau et des
ions minéraux. Il me semble qu'on devrait penser aussi
à l'action métabolique des racines. Par exemple, dans
le pommier, souvent tout le métabolisme de l'azote, la
réduction de l'azote, se fait dans les plus petites
racines, et ceci est extrêmement important. Quand il y
a des mycorhizes également il y a une action sur
l'absorption du phosphore, et peut être aussi sur
l'arrivée de l'oxygène aux racines.

REDLICH: I have noticed great differences in tree
growth response to phosphate, depending on whether the
soil was sterilized.

BARBER: If you heat sterilize soil, you can release an
enormous amount of magnesium, zinc and all sorts of
things. It may not be connected at all with fungi.

BOWEN: We have conducted many studies now on soil sterilization, and found that by far the best method for
laboratory studies is $\gamma$ radiation sterilization. When
we inoculate an irradiated soil with micro-organisms,
the nitrifying bacteria are very slow to multiply so we
get an ammonia fed plant instead of a nitrate fed plant.
We have used this to look at ammonia versus nitrate nutrition by plants in a non-sterile system.

WAISEL: I think it is unfair to look at the roots as
absorbing organs, although certainly they do it; but
we don't have to look at it only as a surface and
relate the yield to the absorbing surface.

    In various water plants which take up all the nu-
trients through the leaves, the roots are still essential for normal growth.  The root tips produce certain
hormones that are essential for the top growth.  So, if
we get two root systems of the same area, one with many
laterals, and one with very few, the plants might show
extremely different growth rates; although the surface
areas are the same.

SAMISH:  We have seen marvellous results today from
mycorrhizas in connection with phosphorus.  I cannot
help thinking that it might be rewarding to extend
this research to heavy metals, elements which are
usually present, but available with difficulty and may
need contact absorption.

BOWEN:  Increased phosphate, magnesium, iron, potassium,
sodium, calcium and boron uptake have been recorded
with endotrophic mycorrhizas.  I believe mycorrhizas
are involved in water uptake also.

MASON:  I would just like to make a comment on the fact
that all the root systems described this morning are
real root systems.  In my section at Summerland Research
Station, Canada, Dr. Stephenson, who is a soil physicist
is working on moisture uptake using root systems, but he
is usually using artificial root systems.  Perhaps artificial root systems might bring more light to bear on
the questions which were expounded this morning.

    The system he is using is one of very small diam-
eter porous ceramic tubes.  These tubes are approximately one millimeter inside diameter, 2 mm outside
diameter, and he is able to put these together in small
plastic boxes so that he can have a set of four tubes,
perhaps separated by one millimeter and then another
set around that again, separated by another one millimeter, so that the effect of different spacing of roots
can be used.

Obviously ceramic tubes are not the sort of thing
that we are interested in for nutrient uptake, but the
idea immediately occurs that perhaps a root made of
synthetic resins in which we can get an ion exchange
surface, might bring some information to bear on the
subjects that we are discussing.

DE WIT:   Such artificial roots would have to be supplied
with a transport mechanism.

BIELESKI:   Ginsburg is using the outer cortex of corn
roots, perfusing the inside, and his looks like being a
very promising way of getting at some of these problems.

WAISEL:   I can't agree about using those semi-artificial
roots, because it is known that if you disturb the intact roots, you get a completely different metabolism,
a different selectivity and a different incorporation
of phosphorus.   If you let tissue age for about 24
hours, you get a completely different incorporation of
phosphorus.

DE WIT:   In view of the conflicting evidence concerning
the location of nitrate reduction in certain plants,
such as the apple - to which relative extent it takes
place in the roots or the leaves - I wonder whether
there exists any basis to think that this may be connected with the presence of mycorrhiza playing an active role in nitrate reduction.

BOWEN:   Fungi have been shown to contain nitrate reductase, but in fact they prefer ammonium ion for their
nutrition.   Presumably they could affect nitrate reduction, but we have no actual evidence.

    If we assume that the extent of the root system is
important, can we increase this root system?

KAFKAFI:   You can't increase the relative root surface
unless you change the whole metabolism.   You can

create different temperatures around the roots and the
tops; but one looks forward to getting different genetic types with roots which might grow faster at the
same temperature.

BARBER:  One way to increase root systems is geneti-
cally.  We have looked at varieties of soybeans, and
we have found some that have twice the root system, in
terms of lengths and areas, of what other varieties do.
We found only a few varieties to have this, and one or
two happened to be the best ones that we had in terms
of yield.

DE WAARD:  We have found that when fertilizers were
placed near the root tips of pepper this plant grew
very well and was completely healthy, and had a production of somewhere about 20-25 kgs per plant, which
had never been shown in that particular crop.  If you
broadcast the fertilizer around the plant, so that the
root has to look for it, then you will never have this
production.  The plant will deteriorate very rapidly,
and the production will be very low.

      I believe, in fact, that this is practical inte-
gration of all the things that Drs. Barber, de Wit and
Kafkafi have told us this morning.  In our studies
water was not limiting, and the soil temperature was
25°C.

BOWEN (Session Leader):  We are reaching a very interesting and exciting phase in the study of soil-plant
relations.  We have heard how we have advanced theor-
etically in simulating soil-plant interactions, and I
am sure that this is going to lead to further ideas
being evolved.  On the other hand, we have people such
as Dr. Barber who are providing us with good techniques
for experimentally testing these ideas. As yet, I don't
really think we know nearly as much as we could on root
growth and root function and the effects of micro-organisms
on these, and I am sure that in the next few years these
are going to be areas of increasing study.

The Volcani Institute of Agricultural Research, at Bet Dagon. Right: Horticulture Building, Left: Agronomy Building. The Volcani Institute is a government institution to which a number of field stations are affiliated. There are a total of 1333 employees of which 299 are scientists.

Auditorium of the Faculty of Agriculture, Rehovot Campus of the Hebrew University of Jerusalem. In 1969/70 it was attended by 669 students (193 post graduate). The academic staff numbers 90 members.